The Future for European Energy Security

Edited by
Curt Gasteyger

St. Martin's Press, New York

First published in the United States of America in 1985

ISBN 0-312-31439-6

Library of Congress Cataloging in Publication Data

Gasteyger, Curt Walter, 1929–
The future for European energy security.
Includes index.
1. Energy policy—Europe. 2. National security.
I. Title.
HD9502.E84G37 1985 333.79'094 85-18398
ISBN 0-312-31439-6

Contents

List of contributors

A. P. Bell	Manager, Corporate Affairs, ESSO Europe Inc., London, UK
David Blair	Research Associate at the Institute of International Relations, University of British Columbia, Canada
Fereidun Fesharaki	Research Associate at the Resource Systems Institute, East–West Center, Honolulu, Hawaii, USA
Curt Gasteyger	Professor at the Graduate Institute of International Studies, and Director of the Programme for Strategic and International Security Studies, Geneva, Switzerland
Hanns Maull	Geschwister-Scholl-Institute for Political Science, University of Munich, FRG
Øystein Noreng	Research Director at the Norwegian School of Management, Oslo, Norway
Antony F. G. Scanlan	Economic Adviser, British Petroleum International Ltd., London, UK
Wendy Schultz	Member of the OPEC Downstream Project, Resource Systems Institute, Honolulu, Hawaii, USA
Ian Smart	Energy Consultant, Britannic House, Moor Lane, London, UK

Acknowledgements

In 1981 the 'Programme for Strategic and International Security Studies' (PSIS) at the Graduate Institute of International Studies in Geneva began a major research project entitled 'Swiss Energy Security in Oil and Gas: The International Dimension'. The research concentrated on the two main areas from which Switzerland and in fact Western Europe as a whole receive a strategically and economically important share of their oil and gas supply, i.e. the Middle East/Gulf region and the Soviet Union. Inevitably, the project acquired a European dimension in which the future of Europe's energy supply as a whole occupied a central place.

It is within this broader context that an international workshop was organized in the spring of 1983 to which eminent specialists on oil and gas from Europe and the United States were invited. Its central theme were the prospects of, and future challenges to, Europe's energy security. A number of papers on various aspects of this topic were commissioned and discussed. The present volume is the result of this meeting.

Our sincere thanks go first to the Swiss National Fund for having financed the entire project including the workshop; second, to the distinguished authors for having taken the time and the trouble of revising and updating their papers, and, last but not least, to Dr Shahram Chubin and Bernard Zen Ruffinen, Director of Research and Research Assistant of the PSIS, respectively, for having so competently carried out the research project and successfully organized the workshop.

Curt Gasteyger

Geneva, August 1985

Chapter 1

Introduction

Curt Gasteyger

Is energy security still an issue in the daily life of Europeans? Considering the way they continue to drive, consume and waste energy the answer is probably 'no'. The tables seem to have turned: it is the oil producing countries that are worrying about stagnating consumption and declining prices, while the consumers, including those in Western Europe are profiting from an 'oil glut'. They do so after several years of anxiety about new crises and higher prices, and profit both from the disarray in OPEC and from the abundance of oil whose end, in 1985, is still over the horizon. It seems to be a period when Europe can relax and allow energy security to sink back to the bottom of its political agenda.

This book attempts to show that such a sanguine attitude is not only premature and misleading but hardly does justice to the complexity of what we call 'energy security'. In other words, its purpose is to analyse the lessons of the recent past, to draw up a balance sheet of Europe's energy situation today and to define the broader contours of its expected evolution to the end of the 1980s. Its central message is that while Europe's dependence on oil from the Middle East has somewhat declined and the diversification of energy supply has made tangible progress, Europe will still be vulnerable to various kinds of challenges and crises of which disruption of supply is only one, and not necessarily the most serious, contingency.

The key aspects of world energy for the rest of this century have been aptly described as, first, the physical sufficiency of energy resources; second, the continuing importance of oil; third, the industrial and developing world's need for imports; fourth, the critical importance of supply from the Middle East; fifth, the impact of new claimants for a share of international oil; sixth, the price of oil; and, seventh, geopolitical constraints, particularly Gulf politics, which now permeate affairs concerning the supply of oil, and, increasingly, of gas.[1]

This enumeration is reassuring in that it emphasizes the sufficiency of energy resources. But the list also serves as a warning: the need for the import of energy resources will not only remain but be increased as there will be more consumers to cater for; the Middle East will retain its critical importance as supplier but may be subject to further political changes and instability, and, finally, abrupt price increases remain a, if not *the*, most important threat for the consumers. All this confirms Ian Smart's opening statement that our notion of 'energy security', despite its popularity, is far from being very clear: it can mean self-sufficiency or independence in terms of supply, predominantly oil supply. But it means also protection against sharp price increases, against restrictions, undesired political pressure or turmoil, or against these eventualities together.

Europe's views or policies are still very much, perhaps too much, shaped by the experience of the 'oil shock' in 1973, and possibly that of 1979. They are therefore one-dimensional, focused too exclusively on one particular scenario at the expense of neglecting several others that, in the mean time, have emerged or may emerge. In short, we must recognize that Europe's energy situation at home and the environment by which it is influenced, have substantially changed since 1973.

Various chapters of this book describe the nature and extent of these changes. In 1983, 48 per cent of Western European primary energy supply came from oil, of which 71 per cent was imported: 21 per cent from coal (18 per cent imported), 15 per cent from gas (21 per cent imported) and 16 per cent from hydraulic, geothermal and nuclear power. In 1985, therefore, imported oil provided only about one-third of Europe's primary energy—a dramatic change on 1973 when its share made up almost 60 per cent of total supply. The causes of this change are well known: they are to be found in a net decrease of consumption, a diversification to other sources and some (fairly modest) measures of conservation.

Both Hanns Maull and A. P. Bell agree that Europe's dependence on oil has declined. There are many indications that this may remain so in future though Bell invites us to be cautious in our forecasts: there have been too many false predictions in the past. Maull in turn stresses that in spite of the welcome reduction in dependence, it is still considerable. Whatever the progress made, he argues, there can be no doubt that Western Europe remains tied to oil supplies from

OPEC regions for an indefinite future. Dependence *per se*, however, is not necessarily a problem. It can be economically advantageous, if it involves supplies that cannot be produced at comparable cost at home, and politically harmless, if it does not imply vulnerability as well. It can be compared with a situation of total vulnerability, which Maull defines it as 'dependence plus', i.e. as a situation in which supplies from abroad might be subject to serious discontinuities which cannot be easily remedied by the importing country, in which the consequent shortfall cannot be easily absorbed or balanced out through alternative supplies and/or painless cutbacks in consumption; in which, then, the cost entailed by an interruption is very high.

In this sense Europe's dependence on oil imports, particularly from as volatile a region as the Middle East, remains a serious problem. Above all the Gulf is, as Fereidun Fesharaki and Wendy Schultz demonstrate, central and will remain so certainly well into the 1990s. According to them, about 50 per cent of potential addition to world oil reserves is likely to come from that area. However, estimates differ and forecasts require some caution: it would seem advisable for policy planning purposes to note that the additions to OPEC oil through secondary recovery are, while substantial, less than previously assumed. On the more reassuring side Fesharaki and Schultz argue that some of the more important oil producing countries—such as Iran and Iraq—will, because of their own needs, pursue an oil production policy that is to a great extent independent of the regime which rules the country. In other words, we can assume that even as drastic a change in the political orientation of a country, as happened in Iran after the fall of the Shah is unlikely to be followed by an equally drastic change in its oil production and export policy.

What can be done to reduce further Europe's dependence on the Middle East? On the whole, the answers are fairly well known. To increase energy production within Europe is one possibility, but A. P. Bell argues that the margin of manoeuvre here is quite small and the main problem is lead time. We know now that almost regardless of the need or the incentive, it takes about ten years from start to finish for a major energy project to come on stream, be it an offshore oil project or a nuclear power plant. Interest in nuclear power has noticeably declined as a result of both a decline in demand for electricity and a widespread opposition to nuclear reactors. On the other hand, the North Sea has turned out to be an important resource base for oil and gas. It thus represents the most attractive source for Europe's greater self-reliance.

Then there is the possibility for further diversification of external supply. Here it should be kept in mind that greater diversification may also imply broader risks. In other words, the criteria used by Maull to define the essence of 'vulnerability', i.e. probability of disruption, possibilities of successful crisis management, and costs of shortfalls—must also be held against the three other import-related energy sectors besides oil, i.e. coal, natural gas and uranium. Neither coal nor uranium present major problems since they come mostly from industrialized countries such as Canada, Australia and the United States.

The situation is different, however, with respect to natural gas. Major imports of natural gas come from Algeria, Libya and the Soviet Union, in addition, of course, to production from Norway and the United Kingdom. None of these extra-European sources is risk-free, though obviously these risks may be of a different nature. Two chapters deal respectively with the Soviet Union and Norway. They are, as will be seen, interlinked. In the 1970s the Soviet Union became the world's main oil producer. As a late-comer the Soviet Union was able to increase production continuously until the end of the 1970s, first in the Volga–Ural area and then, as from the 1960s, in Western Siberia. According to Antony Scanlan 'virtually the whole of the subsequent development of all-Union production centred on this latter region ever since. This is true for Soviet oil and even more so for Soviet gas for which the reserve base is much greater than it is for oil'. But this era of uninterrupted growth of production is coming to an end, certainly for oil. Traditional oil fields are being depleted, new ones are either much more distant or technically more difficult to exploit, or both. Even with gas, as Scanlan points out, the first indicators of planned targets for the 1986–90 Five Year Plan anticipate slower growth though there are still many potentially fertile basins awaiting development.

What must interest us here is the extent to which the Soviet Union is, and will remain, an important supplier for Europe. With Japan being only a very minor customer, the vast bulk of Soviet exports for energy (about 90 per cent) goes either to Eastern or to Western Europe in 1985. To be sure, Eastern Europe will have to rely on

Soviet exports indefinitely. So Moscow's flexibility with regard to non-communist clients is substantially limited. Soviet energy supplies to Western Europe comprise about 10 per cent of the latter's demand. They constitute the greatest part of Comecon's total energy exports today and in the foreseeable future. This means two things: first, while significant, Western Europe's dependence on such imports from the communist world remains strategically 'controllable'; second, the concentration of Soviet energy exports outside the Comecon area is clearly on Western Europe. From this we can deduce a tangible Soviet interest in a continuation of this relationship. Hence there is a kind of mutual dependence neither of which would seem to cause the kind of vulnerability that Europe still fears *vis-a-vis* the Middle East.

An important reason for this relates to two other suppliers of gas and/or oil, i.e. Britain and Norway. The latter is obviously more important as the chapter by Øystein Noreng demonstrates. The North Sea gas reserves are indigenous to Western Europe and secure. Therefore they represent a politically attractive source of supply. Their main competitors are, in terms of cost, Algeria and the Soviet Union. Any increase of gas exports to Europe from Algeria would, however, necessitate transportation by Liquid Natural Gas (LNG) tankers or the construction of another pipeline in addition to that going to Italy. Either alternative would raise costs considerably. The Soviet Union, on the other hand, has a large reserve base. Her exports could be increased at low marginal cost at least up to a certain limit. This is where Noreng sees competition with Norwegian gas. Both Norway and the Soviet Union have an interest in knowing what share of the European gas market they can hope for—and at what price. In a sense, they are rivals and partners at the same time: while obviously competing for contracts on the European market neither has an interest in 'flooding' it with gas and, as a consequence, risking a depression in prices. For Norway, Noreng argues, the key issue today is whether to expand gas output over the next ten to fifteen years and at what price or, alternatively, whether to opt for an accelerated development of oil. Much of the answer to these questions depends on Western Europe. Simply put, the latter will have to make a choice between politically attractive Norwegian gas or economically more interesting Soviet gas.

Faced with such a choice, Noreng argues in favour of a long-term strategy for the following reasons: with more gas reserves at hand (thanks to recent discoveries), Norway can afford to give higher priority to volumes and market shares than to high prices. To the extent that its gas takes a larger share of the market, Western Europe's bargaining position *vis-à-vis* the Soviet Union will improve. Provided that Norway could then lower its prices, such strategy would, on the one hand enhance the stability of gas supply from a political ally and reduce dependence on a potentially hostile Soviet Union on the other.

When appraising the various causes of threats to Europe's energy security the authors agree that there are several others which are more important than a lack of resources. Such a lack is not the real problem, at least not in the foreseeable future. What is more worrying is, first, the degree of dependence, including that on Soviet gas, which is still high; second, the emerging competition in the field of petrochemicals, and, third, the possibility of a sudden disruption in the oil supply with the almost inevitable surge of prices. The latter possibility remains no doubt one of the major sources of European vulnerability in energy security.

Several questions arise out of this. They have to do with the way (concerted or not) such crises are handled by both the suppliers and the consumers; the rapidity and effectiveness–as Blair points out–with which the International Energy Agency can intervene; the role the United States will play; the behaviour of the various market forces; and, last but not least, the reaction of the general public. Views obviously differ as to the extent to which governments should intervene in such a critical situation and what the–no doubt reduced–role of the multinational companies can and will be.

Maull concludes his chapter by stating that 'oil is too important a commodity to be left to market forces alone'–a statement which expresses a legitimate concern but does not answer the more delicate questions as to the extent and nature of state intervention and, more complex still, the coordination of such intervention with that of other countries on an international or European level. The latter issue is of concern to those authors who wonder whether, when and how European countries will and can develop a more concerted and crisis-resistant energy policy. Such a policy would seem indispensable if the uncertainties of the oil and gas market become greater and protection against abrupt price surges again becomes topical.

Indeed, another kind of agreement may become necessary in the not too distant future in the petrochemical industry. Fesharaki and Schultz rightly draw attention to the growing challenge which may come from supplier countries in downstream operations in general, and the petrochemical industry in particular. Many OPEC countries have made a serious move into this field. They have begun to industrialize with the result that in 1985 petrochemical plants in Saudi Arabia, Mexico, Kuwait and other 'newcomer' countries make up some 10 per cent of world capacity. Hence growing fears in both Western industrialized and newly industrializing countries that many existing petrochemical plants may risk at least partly redundancies in five to eight years' time as supply begins to exceed demand and prices increasingly come under pressure. It is a matter of both speculation and concern as to whether such pressure will hit the Western or the newly industrializing oil-producing countries more.

The main message to Europe of this and other possible developments is, in the words of Ian Smart, that energy supply will never be secure 'because insecurity is in the nature of energy and the world'. He continues: 'Apart from the security aspect of economic costs, risks of accident, industrial conflict and environmental prohibition always hang over indigenous supply'—however autonomous such supply may become over the years. In other words, the search for complete energy security will remain as vain as that for complete military security. It may even become counter-productive. There are, and will be, forms of insecurity and challenges to security that are either not foreseeable or simply not controllable. All the authors, either explicitly or implicitly, therefore agree that what matters is to understand the general trends of change and challenge, and to strike the best possible balance between the risks of traditional sources of supply on the one hand, and the costs of alternative ways on the other.

To what extent this can be done successfully depends not only on the policies of one single country alone, however shrewd and farsighted, but ultimately on the degree of consultation, coordination and cooperation amongst the European countries as a whole and, as Blair points out, on the confidence of each of these countries in the reliability and efficiency of such cooperation. In future crises, the energy security of each of them, and indeed of the Western countries in general, will depend therefore very much on the response of the others. It sounds like an easy answer to a complex question. But if past experience and a serious examination of likely developments in the future, as provided by this book, is any guide, this is the main conclusion that Europe ought to draw.

Notes

1. Cf. Conant, Melvin A., *The Geopolitics of Oil*, Center for International Security Studies, University of Pittsburgh. *Occasional Paper*, No. 1981/2 (June 1981), Pittsburgh.

Chapter 2

Europe's energy situation

Hanns W. Maull

Europe's energy equation 1973 and 1980; from the first to the second oil shock

From the comparative data of the European* energy position in 1973 and 1983 (presented in Table 2.1), a number of rapid conclusions can be drawn:

- Total energy requirements in Western Europe actually declined between 1973 and 1983 in spite of a considerable expansion of European economies.
- Energy production within Western Europe rose significantly—with oil playing by far the most important role in this, followed by nuclear power and natural gas.
- Demand for oil declined markedly in spite of much higher indigeneous production. As a result, demand for oil imports contracted even more sharply. The share of oil in total energy consumption was reduced.
- Overall dependence on imported oil was reduced quite significantly, although this was to some extent due to a shift towards imported natural gas and coal; in those two energy sectors import levels rose modestly. Still, higher indigeneous supplies and, above all, the decline in energy demand played the key roles in reducing overall import dependence.

It is instructive to compare those changes in the European energy position between 1973 and 1983 with the 'conventional wisdom' of energy experts after the first oil shock in 1974–5. Practically nobody amongst the established energy economists thought that energy consumption could remain much the same in spite of substantial economic growth; those (such as Amory Lovins) who did argue that expectations about future energy requirements were wildly exaggerated were considered exotic birds—interesting but useless.

* 'Europe' or 'Western Europe' is used interchangeably for the OECD members in Europe.

Most experts also assumed that higher oil prices would bring forth very vigorous development of alternative sources of energy. Nuclear power, in particular, was seen as a major contributor to changes in energy consumption patterns, but there were also great hopes for additional oil, natural gas and coal production within the OECD region, and within Western Europe.[1] Others, however, cautioned that, at least for the OECD region, taken together, additional supplies of oil would only be sufficient to compensate for the US oil production decline in the lower forty-eight states, that the development of alternative sources such as coal, nuclear power and synthetic fuels could well be slowed by their social and economic costs and risks. As it turned out, those sceptics were basically correct—the development of domestic production of energy in Western European countries has been less successful than hoped and expected in the mid-1970s; the great strides suggested in Table 2.1 are somewhat deceptive in that they include the period of opening up a major new energy province in Western Europe, the North Sea, with its oil and gas. In the future, such a bonus could only come from a massive expansion of nuclear energy.

Table 2.1 Western Europe's energy balance, 1973 and 1983 (mtoe)

	Production		Imports*		Total requirements		Import dependence	
	1973	1983	1973	1983	1973	1983	1973	1983
Oil	22.5	171.3	767.2	373.9	748.9	585.7	102%	64%
Solid fuels	229†	214.9	36†	47.5	265†	183.5	14%	18%
Natural gas	117.7	144.8	12.2	38.7	129.9	183.5	10%	21%
Nuclear power	16.6	79.5	–	–	16.6	79.5	–	–
Other	91.9	110.7	–	–	91.9	110.7	–	–
Total	477.7	721.2	815.4	460.1	1,240.9	1,221.8	66%	38%

* Actual for oil, statistical difference between production and consumption for other fuels. Note that imports and production of oil in 1973 exceeded requirements but were lower in 1983, reflecting stock build and stock draw, respectively. The effect exaggerates the decline in oil import dependence by about 12 per cent.

† OECD data, which differ slightly from BP figures.

Sources: *BP Statistical Review of World Energy*; OECD Energy Balances.

All the established experts were wrong, however, as far as the evolution of energy demand was concerned. It was the relative stagnation of energy demand over the past decade that made major progresses towards reducing energy import dependence possible. This very slow growth of energy consumption pushed the somewhat dubious correlation between economic growth and energy consumption growth from 1 (i.e. 1 per cent economic growth required a 1 per cent increase in energy production) in the period 1960 to 1973 to 0.1 (i.e. with average economic growth of 2.3 per cent annually in the OECD, energy demand grew by only 0.2 per cent per annum) for the years from 1973 to 1981.[2] In fact, it is now virtually impossible to project the future relationship between economic growth and energy demand. More meaningful is a sectoral analysis of energy consumption, which can take into consideration structural changes in European economies (for instance, the change from industry to services, and from 'old', energy-intensive industries such as steel to 'new' industries such as electronics and communication), cyclical aspects of energy consumption (economic recessions affect energy-intensive industries particularly seriously and thus tend to compress energy consumption beyond what would be implied by the underlying changes in consumption patterns), and the adjustment to higher energy prices in the shape of greater energy efficiency.

All three of those elements (structural change, cyclical aspects and effects of higher prices) have been at work during the period under consideration—although the longer-term implications of the second oil shock were probably still to show their full impact in 1983. Thus, the consumption patterns in that year can reasonably be taken as a picture of adjustment to the first oil shock plus the cyclical downturn triggered by the second oil crisis. Much more difficult to decide, however, is the contribution made by each factor in reducing energy requirements so much below what had been envisaged.

Beyond this virtual stagnation of energy consumption, the reduction of oil import dependence was achieved through a shift in energy consumption patterns, away from oil towards natural gas, nuclear fuel and, to a small extent, coal. Overall, this shift was quite marked. It showed a decline of oil's share from about 60 per cent to about 48 per cent of total primary energy requirements between 1973 and 1983. But this shift could not be achieved fully through higher indigenous production of those alternative sources of energy but had to draw on increased imports of natural gas, coal (and uranium).

Table 2.2 completes the picture of Europe's energy development since 1973 with a look at some individual countries. This set of figures shows that actual developments have differed very much from country to country. Thus total requirements in the United Kingdom declined quite strongly, while domestic production of oil and gas surged ahead. This left the country with a significant energy surplus in 1983. France and the Federal Republic of Germany (FRG) both show marginal changes in total energy consumption and much steeper reductions in oil consumption. This reduction was more pronounced in France, where the shift towards (imported) natural gas and coal as well as to nuclear power was vigorously pushed by the government. In contrast, the FRG traditionally relies more on market mechanisms. Overall, however, the differences in performance were less than one might have expected, given the spectacular successes of the French nuclear programme and the difficulties encountered by nuclear power in Germany. All told, the FRG in 1981 still was in a more comfortable energy position–its large coal production, as well as sizeable supplies of domestic natural gas more than compensated for French advances in nuclear energy–all the more so as the poor energy resource endowment of France was painfully felt in declining production of coal and natural gas. The case of Switzerland illustrates the limited possibilities of resource-poor countries–although this country is fortunate in having fairly ample hydropower.

Europe's oil import dependence: patterns of supply and their implications

Western Europe has thus been able to reduce its overall energy import dependence quite significantly–from 62 per cent to 41 per cent between 1973 and 1983, if one excludes stock variations. The share of imported oil in total energy requirements declined even more: from 58 per cent to 34 per cent. Moreover, the share of OPEC suppliers in total European imports of oil was 81 per cent in 1980–but only 75.5 per cent in 1981. Dependence on OPEC supplies, expressed as a percentage of total energy requirements, fell below 30 per cent in 1983. The regional distribution of sources of supply for

Table 2.2 Individual countries' energy balances, 1973 and 1981 (mtoe)

	Production		Imports		Total requirements	
	1973	1981	1973	1981	1973	1981
Federal Republic of Germany						
Oil	6.7	5.1	146.5	111.5	153.2	116.6
Solid fuels	92.8	93.0	−10.4	−4.3	82.4	88.7
Natural gas	15.4	14.9	12.2	26.8	27.6	41.7
Nuclear power	2.7	12.7	–	–	2.7	12.7
Other	3.5	4.7	–	–	3.5	4.7
Total	121.1	130.5	148.3	134.0	266.4	264.5
Great Britain						
Oil	0.4	91.4	116.5	−18.5	116.9	72.9
Solid Fuels	78.5	73.2	−0.9	−3.7	77.6	69.5
Natural gas	25.0	31.8	0.7	9.8	25.7	41.6
Nuclear power	7.2	9.5	–	–	7.2	9.5
Other	1.2	1.4	–	–	1.2	1.4
Total	112.3	207.3	116.3	−12.4	224.0	194.9
France						
Oil	2.1	2.5	129.6	88.2	117.3	90.7
Solid fuels	19.4	15.5	10.0	16.6	30.5	32.1
Natural gas	7.0	6.7	7.7	18.0	15.0	21.9
Nuclear power Other	3.1 10.5	22.1 15.8	−0.2	−1.0	12.5	36.8
Total	42.2	62.6	147.1	121.7	175.3	184.3
Switzerland						
Oil	–	–	15.1	11.7	15.1	11.7
Solid fuels	0.2	0.3	0.2	0.7	0.4	1.0
Natural gas	–	–	0.2	1.0	0.2	1.0
Nuclear power	1.5	3.6	–	–	1.5	3.6
Other	6.8	8.6	–	–	6.8	8.6
Total	8.6	12.4	15.5	13.4	23.9	25.9

Sources: OECD Energy Statistics; IEA Annual Reviews; Comité Professionnel du Pétrole.

1983 showed the following picture: of total European crude oil imports of 430 million tons some 169.2 or over 39 per cent came from the Gulf region, a further 28 per cent (120.5 million tons) from Africa, 51.1 million tons or 12 per cent from the Western hemisphere, and 89.2 million tons or about 21 per cent from others, mostly the USSR and Eastern Europe, with 87.8 million tons.[3]

In other words, levels of dependence are still quite considerable. Whatever the progress made, there can be no doubt that Western Europe remains tied to oil supplies from OPEC regions. Dependence *per se*, however, is not necessarily a problem–it can be economically advantageous (if it involves supplies which cannot be produced at comparable cost at home) and politically harmless, if it does not imply vulnerability, as well. Vulnerability might be defined as 'dependence plus', i.e. as a situation in which supplies from abroad might be subject to serious discontinuities which cannot be easily removed by the importing country; in which the consequent shortfall cannot be easily absorbed or balanced out through alternative supplies and/or painless cutbacks in consumption; in which, then, the cost entailed by the interruption is very high.

Oil import dependence in that sense clearly still is a problem. The volatility of Middle East politics has produced numerous explosions with consequences for world oil flows–in 1948, at the time of the foundation of the state of Israel; in 1951–3 during the Mossadegh oil nationalizations in Iran; in 1956 and 1967 as a result of Israeli–Arab wars; in 1973, again in connection with an Israeli–Arab war, when the 'oil weapon' was used; and in 1978–9 as a result of the revolution in Iran. There can be little doubt that during the next two decades more surprises are in store for world oil markets. They are likely to come from the Middle East–but they need not be confined to this area. A study by the US Department of Energy thought a 3 million barrels per day (mb/d) interruption of supplies had a probability of 50–95 per cent over a decade, depending on the assumptions made; for a supply shortfall of 10 mb/d, the study sees a 30 per cent probability over a ten-year period as the most likely guess.[4] Thus, supply disturbances must be considered quite likely. This fulfils the first condition of our definition of vulnerability, as under most scenarios it would not be easy for the importers to remove the cause of the disturbances.

The second condition stipulates that crisis management might not be able to contain the costs of adjustment to the shortfall. Demand for oil is not very elastic in the short run: in other words, consumption cannot be lowered very much without serious economic and social inconveniences and losses. Alternative sources of supply—either in the form of mobilized standby production from oil exporters unaffected by the cause of the problem, or through fuel-switching in dual-capable boilers (for instance in power stations)—would not necessarily be able to replace much of the shortfall. Spare oil production capacity will probably be largely concentrated in the Gulf. It is therefore not easily available, and the possibilities of fuel-switching are, in any case, limited. Stockpile drawdowns and rationing programmes are then the next lines of defence. But they are problematical instruments of protection.

Thus, crisis management capacities (the command over technical, physical and managerial resources needed to facilitate smooth adjustment in situations of supply shortfall) might not be sufficient. One could also demonstrate this point historically: during the past decade, there were two major supply crises which were managed relatively badly. There was only one (in 1980–1, resulting from the Iran–Iraq war) where crisis management, for a variety of reasons, was quite successful. This should caution us against over-confidence about our ability to do well in the next crisis.

The third condition for establishing vulnerability has to do with overall costs. They have been massive even in the case of relatively small supply shortfalls. The shortfalls associated with the Iranian revolution and its aftermath never reached the levels needed to trigger the International Energy Agency Emergency Allocation system, OPEC production totals during 1979 for every single month exceeded figures in 1978 (other OPEC suppliers filled the gap left by Iranian production cutbacks reasonably well). Yet this small shortfall, combined with stock movements which procyclically accentuated pressure on supplies, was sufficient to cause a second oil price explosion which, in turn, led to a loss of about 5 per cent of OECD real income in 1980, and close to 8 per cent in 1981. The total sum approached $1,000 bn. or $1,250 bn per head for all industrialized countries.[5] Those figures represent necessarily only rough approximations. Still, they are conservative in assessing the economic consequences of the price explosion. They do not even take into consideration less tangible political costs such as the weakening of the fabric of international cooperation.

These considerations try to define more rigidly what does and what does not constitute a problem with respect to energy import dependence. At the same time they make clear why so far there has been little attention paid to import dependence with respect to coal or natural gas. The levels of dependence implied in the data in Tables 2.1 and 2.2 simply do not constitute vulnerability. Experiences with losses of Algerian gas and Polish coal have shown that supplies of those fuels could well be disrupted, too, but they also demonstrate that so far any supply loss could easily be substituted and/or absorbed.[6]

From the present into the future: diversified energy structures, broader risks

This exclusive consideration of energy security in terms of oil security will no longer be appropriate in the future. First, all forecasts and projections about the evolution of Western European energy demand to the year 2000 assume significantly higher levels of import dependence for coal and natural gas. Second, nuclear power, generally credited with good chances for an expanded role, relies on uranium, which will also have to be imported by most European countries.

Table 2.3 presents the projections by the International Energy Agency's (IEA) 1982 *World Energy Outlook*. Two scenarios are considered: the central assumptions are a decline in the real price of oil to 1985 and then a constant real price in the high demand scenario. Economic growth averages 3.2 per cent from 1985 onwards in this scenario. The low energy demand scenario assumes slightly lower economic growth rates after 1984, and a 3 per cent real increase per annum in oil prices from 1984 onwards. The IEA projections are compared in the table with estimates by Exxon[7] so as to provide a differently coloured glimpse in the crystal ball. Exxon's projections are considerably lower in terms of overall energy demand, and consistently less bullish about coal and nuclear power. To a large extent, these differences can probably be explained by the lower growth rates projected by Exxon.

Table 2.3 Western Europe: the energy outlook according to IEA and Exxon (mtoe)

	Oil		Coal		Natural gas		Nuclear power		Other		Total	
	A	B	A	B	A	B	A	B	A	B	A	B
1990												
Production	128–133	160	220–236	202	160–170	150	199–206	167	111–120	139	843–850	820
Imports	408–450	402	137–149	109	90–100	70	–	–	–	–	635–699	570
Total	541–578	562	367–385	311	260	220	199–206	167	111–120	139	1478–1549	1390
(Import dependence, %)	(75–78)	(72)	(37–38)	(35)	(35–38)	(32)	–	–	–	–	(43–45)	(41)
2000												
Production	109–149	144	290–300	215	125–150	137–162	266–296	240	170–190	144	1020–1025	930
Imports	359–641	417	271–279	184	130–140	88–133	–	–	–	–	778–1042	673
Total	508–750	561	569–57	399	255–290	250	266–296	240	107–190	144	1803–2062	1603
(Import dependence, %)	(71–85)	(74)	(48–49)	(46)	(48–51)	(35–45)	–	–	–	–	(43–50)	(42)

Sources: A: IEA, *World Energy Outlook* (1982); B: Exxon, *Energy in Europe* (December 1982).

The sets of estimates are sufficiently in agreement with each other to allow some broad conclusions:

- a more or less stable or slightly declining production of oil and gas in Western Europe;
- probably not much increase in solid fuel production either (although the IEA sees some expansion by 2000, this seems to be coloured by British (and German) official optimism, which so far does not look very convincing);
- a strong expansion of nuclear power;
- declining oil imports throughout the period, although the IEA projections show a wide margin of uncertainty towards the turn of the century, indicating the possibility of a trend reversal;
- rising levels of imports of natural gas, coal and (although this does not appear in the projections itself) of uranium.

The share of oil in total energy consumption will continue to decline, and with it dependence on imported oil in terms of its share in total energy consumption. This is, however, *a priori* not a very useful indicator because it assumes perfect interchangeability of fuels. In reality, the possibilities of switching from one fuel to another at short notice, and even with some lead times, are limited. A more meaningful indicator of independence is, therefore,the percentage of imports in one particular energy sector. It is perhaps more interesting to note that dependence on oil imports in those terms will probably not decline very much more, while that on natural gas, coal and uranium imports will increase. Western Europe is thus faced with a diversification of import dependencies.

This raises a number of interesting new issues. The first has already come up extensively in connection with the Western European gas pipeline deal with the Soviet Union. It concerns the impact of a diversification of risks on overall energy security. The Western European argument has been that such a diversification has a positive impact on overall risks; the United States' response sometimes appeared to imply the opposite. In fact, there is probably no general answer to this question. On balance, one might well conclude that it does not make much difference whether 70 per cent or 80 per cent of total oil requirements are imported—as long as the additional imports (resulting from lesser efforts at diversification) do not themselves make a supply disruption more likely (by putting pressure on available supplies) or less manageable (by absorbing spare capacity which could be used in a crisis).

A second, more central question relates to the issue of vulnerability with regard to energy sources other than oil. The criteria used in this chapter–probability of disruption, possibilities of successful crisis management, and costs of shortfalls–have thus to be held against all four major import-related energy sectors, that is to say, oil, coal, natural gas and uranium. We can discard problems with regard to coal and uranium fairly easily: most imports will come from industrialized countries such as Canada, Australia and the United States; possibilities of adjusting to shortages are ample, and–with the exception of South Africa, another major future supplier of coal and uranium–the causes of possible shortfalls could probably be removed without great difficulties (they will in all likelihood consist either of labour disputes or of technical failures or natural disasters, in themselves quite unlikely to reach major proportions).

The situation is different with respect to natural gas. There, imports will come from Algeria, Libya and the Soviet Union and, in addition, perhaps from West Africa (Nigeria, Cameroon) and the Gulf. In other words, the main supplies will be from OPEC and the Soviet Union. The latter and Algeria will–even if only firm contracts are counted–reach market shares of around 30 per cent each in several Western European natural gas markets. The possibility of supply disruptions–either for reasons of economic or political coercion or because of turmoil in producer regions–cannot be disregarded. Thus, the first condition for vulnerability is met.

At the same time, the market shares reached by the Soviet Union and Algeria, separately and in conjunction, are sizeable enough to pose serious potential crisis management problems. To be sure, there are considerable flexibilities built into the Western European natural gas system. They consist of:

- a large and well-integrated European pipeline grid. It allows for shortfalls to be distributed widely and to roll in additional supplies from within Western Europe;
- considerable surge capacity in the Netherlands and, to a smaller extent (and depending on circumstances at the time of disruption) in the FRG. There might also be some flexibility in stepping up production in Norway;

- storage: most storage capacity at present is needed to cope with seasonal fluctuations in demand, but capacities are expanded and this will yield a strategic reserve in some countries (France, Italy);
- interruptible contracts: a sizeable share of natural gas is marketed in Western Europe on the basis of contracts which allow suppliers to disconnect customers for limited periods. Again, this instrument is needed to balance supply and seasonally varying demand. But interruptible contracts imply the existence of dual-fired burners. They thus offer possibilities of fuel-switching even for prolonged periods—provided alternative sources are available. Most dual burners, however, are equipped for the combination gas/oil. This would not be all that helpful if both oil and natural gas were affected by supply disruptions.

The flexibility of the Western European natural gas system is expanding constantly—but import dependence will also grow in the late 1980s and 1990s. The European Commission has probed scenarios for supply shortfalls of 10 per cent and 25 per cent for a six-month period during the winter season of 1990. It concluded that there would be no major difficulties.[8] This result seems somewhat optimistic. Methodologically, one might object that it does not look at the regional (as opposed to the national) implications of such shortages; and a six-month limitation might not be the worst conceivable scenario. Thus, even in 1990, some question marks remain. The problems could well increase as surge capacity in the Netherlands diminishes and import dependence grows further. The second condition for the existence of vulnerability thus must also be considered as present—at least in the 1990s, and as long as crisis management flexibility expands only along lines apparent at present.

The last criterion for vulnerability is the cost of disruptions after the easy options outlined above are exhausted. This cost has two aspects; possible negative implications of large price increases, and of disruptions in production and well-being of citizens. The first point will probably not be relevant with natural gas. Total natural gas exports of the Soviet Union to Western Europe will net some $12 bn. in 1990 and 2000 (in constant dollars).[9] The contract structure for natural gas exports is much more rigid than that for oil, and there are (almost) no spot sales. Thus, dramatic price changes are less likely than in the case of oil. Even if they should occur, their impact on the balance of payment, on inflation and hence on economic policies would be much less dramatic than in the case of oil price explosions.

The impact of supply shortfalls beyond interruptible contracts would largely fall on the residential/commercial sector of gas consumption, which will expand to about half of total demand in Western Europe. Industrial demand outside interruptible contracts is not easily substitutable. This sector would therefore probably have to be protected so as to prevent losses of industrial production. Rationing of supplies to households appears to pose difficult problems. If it could be implemented the effect would be some inconvenience (lower room temperatures) rather than serious losses of material production and welfare. Governments might be reluctant to put this burden on citizens, but in a serious crisis, it should be politically possible. Thus, by this third criterion, vulnerability certainly seems much less dramatic than in the case of oil.

Oil: still the Achilles' heel of energy security

Energy security concerns will thus broaden over the next few years but oil will remain at their core. The simplest reason for this proposition is volume: even by the year 2000, Western European oil imports will still vastly exceed (in energy equivalence) imports of other fuels. This has, of course, implications for economic security. There is the balance of payments effect of price increases. Together with inflationary pressures they could pose major problems for economic management in themselves, even without a significant shortfall of supplies. Whether the fabric of international economic cooperation in trade and finance could survive another oil price shock such as the one in 1979–80 intact, must be in doubt—particularly if such a price explosion should occur within the next few years at a time when the world economy is still trying to adjust to the previous changes. Note that smaller oil price percentage increases could in the future produce effects in terms of their relation to Gross National Product similar to the dramatic jumps of 1973–4 and 1979–80. Note, too, that (assuming stable or increasing oil price trends) oil consumption will tend to become concentrated in sectors where substitution would be difficult or impossible, while the 'fat' of relatively dispensable oil consumption will have been reduced by efforts to enhance energy efficiency. The decline in oil's position in the energy equation could thus be partly balanced by its heightened importance.

A second reason for assuming a continuing key role for oil in security matters has to do with the overlapping of the oil with the natural gas markets. Some of the key suppliers are identical (OPEC countries). It is possible to conceive of political scenarios of oil supply disruptions from the Gulf, which would involve both superpowers. Hence this would link the question of Soviet natural gas supplies to Western Europe to such an oil crisis. The two energy markets cannot be entirely separated politically.

Oil supply disruptions would thus continue to pose extraordinary problems of adjustment both in terms of absorbing a serious shortfall of supplies and of managing the implications of possible large price increases. Moreover, they could spill over into natural gas supply risks. Given future levels of Western European oil import dependence a large shortfall could be absorbed only at high economic, social and political costs. The second and third conditions for establishing vulnerability will therefore continue to be highly relevant. There can be no doubt about the possibility, indeed probability, of other supply crises.

This last point merits some further discussion. The future pattern of oil supplies to Western Europe is, of course, much more difficult to project than that for natural gas. Still, it seems reasonable to assume that Western European oil imports might increasingly come from the Middle East. Table 2.4 compares two sets of estimates about OPEC export potential in 1990 and 2000. They both show a similar trend: OPEC export capacities will increasingly be concentrated in the Gulf. In 2000, the share of Gulf countries in total OPEC exports could reach, according to those projections, between 79 and 89 per cent.

There are several good reasons for assuming that this trend will indeed materialize–size of reserves, population, and the cost of installing additional capacities. A number of non-Gulf exporters will increasingly be hard pressed to maintain production and to meet rising domestic energy demand. There are, of course different policy options and mixes available for those countries: they might either phase out oil exports or squeeze domestic demand to protect their export earning capacities. But, on the whole, their differences in

Table 2.4 Future export capacities of OPEC countries (mb/d)

	Production					Domestic Consumption					Exports				
	1980	1990		2000		1980	1990		2000		1980	1990		2000	
		A	B	A	B		A	B	A	B		A	B	A	B
Saudi Arabia	10.3	3.4	11.0	9.4	12.1	0.6	1.1	1.1	1.7	1.4	9.7	8.3	9.9	7.7	10.8
Kuwait	1.8	1.5	1.7	1.2	2.2	0.1	0.2	0.06	0.2	0.08	1.7	1.3	1.6	1.0	2.1
UAE	1.7	2.0	2.0	1.5	2.0	0.1	0.2	0.2	0.2	0.3	1.6	1.8	1.8	1.3	1.7
Qatar	0.5	0.3	0.2	0.2	0.1	0.01	0.02	0.02	0.02	0.02	0.5	0.3	0.2	0.2	0.1
Iraq	2.7	3.5	3.5	3.5	4.0	0.2	0.4	0.3	0.7	0.6	2.5	3.1	3.2	2.8	3.4
Iran	1.5	3.0	3.0	3.0	3.0	0.6	0.7	0.6	1.5	0.9	0.9	2.3	2.4	1.5	2.1
Subtotal	18.5	19.7	21.4	18.8	23.4	1.6	2.6	2.3	4.3	3.3	16.9	17.1	19.1	14.5	20.2
Libya	1.8	1.8	1.5	1.7	1.5	0.7	0.2	0.2	0.3	0.2	1.7	1.6	1.3	1.4	1.3
Algeria	1.1	0.9	0.7	0.7	0.4	0.1	0.3	0.2	0.4	0.3	1.0	0.6	0.5	0.3	0.1
Indonesia	1.6	1.4	0.9	1.2	1.5	0.4	0.9	0.9	1.2	1.5	1.2	0.5	—	—	—
Nigeria	2.1	1.8	1.5	1.5	1.0	0.2	0.5	0.5	1.0	1.0	1.9	1.3	0.9	0.5	—
Venezuela	2.2	2.2	2.0	3.0	2.0	0.4	0.7	0.7	1.2	1.1	1.8	1.5	1.3	1.8	0.9
Total OPEC*	27.6	28.1	28.0	27.1	30.1	2.9	5.5	5.0	8.7	7.5	24.6	22.6	23.1	18.4	22.6
Total est. production capacity	35.2	29.1–34.1		27.8–33.5											

* Including Gabon and Ecuador.
Sources: A: IEA, *World Energy Outlook* (1982); B: N.Ait-Laoussine.

terms of distribution of OPEC exports would not be great. The increasing concentration of export capacities in the Gulf region will undoubtedly enhance energy security risks: the probability of some form of turmoil in the region affecting oil supplies must be considered high, and the volumes involved could easily become very substantial. Moreover, any crisis in the Gulf is bound to be followed very closely by both superpowers. Thus, there is always a possibility of a major world crisis.

Trend projections such as these about the growing concentration of export capacity in the Gulf have always to be considered in the context of the real oil market with its cyclical ups and downs. Thus, at any given moment in the future, the *actual* share of the Gulf might well be lower—witness developments since 1980. Given the revenue requirements of Gulf and non-Gulf exporters, a reasonable distribution of export markets would involve full utilization of capacity in the non-Gulf regions, with the Gulf countries playing the role of swing suppliers. Whether this assumption materializes will depend on the degree of cooperation or conflict within OPEC. In this scenario, however, spare capacity and thus a large element of crisis management flexibility would still be concentrated in the Gulf. Political events in the Gulf which would affect this spare capacity (say, a new regime in Saudi Arabia with a 5 mb/d production ceiling) would in themselves be sufficient to transform the world oil situation.

'Ceteris non Paribus': the problems of future market instability

The political risk element in the present and future oil equation is one crucial reason for a cautious use of projections. Political events could still transform the oil market over night. They could drastically change prices and even longer-term price trends and expectations. There is, however, a second, perhaps less well appreciated element of uncertainty: changing oil market structures. The changing role of the oil majors from producers to buyers, traders and refiners of oil; the proliferation of actors in the oil market; the declining importance of long-term contracts and the increasing role of spot sales (said to have reached up to 40 per cent of total trade in the early 1980s), and the growing importance of stock movements (which tend to exacerbate the cyclical fluctations of demand) as well as the accentuation of OPEC's role as the marginal supplier suffering from extreme fluctuations in demand for its exports. All this tends to favour market instability and large price fluctuations. To this has to be added the growing fragmentation of the political environment of the oil market: the bitter conflict between revolutionary Iran and conservative Saudi Arabia and secular Iraq; the weakening of United States' influence in the Middle East; and the continuing deadlock of the Israeli–Arab conflict.

The international oil market will thus probably be more fragile and subject to fairly pronounced ups and downs in the future. This will be particularly so if efforts to build a new, viable structure of stability and control fail. Prospects for such a successful structuring, be it by OPEC, by the IEA countries, or through joint efforts, appear rather dim at present. This volatility expresses itself in downward pressure on prices. There is as yet great uncertainty about both the price levels which will materialize, and the longer-term price trends.

I do not want to pursue this further by adding my own guess about the future because it seems wise to start from this assumption of uncertainty. Oil prices could, in fact, be as low as $15 per barrel (pb) or as high as $50 pb in 1986—and under each scenario they might well be again entirely different in 1987. What are the implications of this proposition of uncertainty for Western European energy security?

A first issue to be considered concerns efforts to diversify away from OPEC imports, and to use oil more efficiently. While there are probably good reasons to assume that improvements in energy efficiency will continue, uncertainty about the future and lower present price levels could well discourage diversification. The problem here might be less on the supply side (up to now, most projects cancelled were probably only marginally profitable anyway) than on the demand side. Consumers might well be reluctant to invest in alternative energy installations. The lack of demand could then feed back into lower investment in alternative supplies. This seems to be happening in the case of coal. The net result of this would be higher levels of dependence on oil imports than might be possible. Moreover, if we take into consideration what has been said before about growing market volatility and stock movements, there would appear to be greater potential for sudden surges in oil demand. They might quickly develop into supply bottlenecks.

Secondly, do falling oil prices and market instability affect the probability of supply disruptions and price jumps? On balance, I would argue that they do. The fairly drastic fall of oil income to OPEC from $275 bn. in 1980 to perhaps $160 bn. in 1983[10] has already created economic and political problems in some high absorbers. Squeezed by inflation, the revolution of rising expectations and a declining ability to meet demands, the social and political stability of producer governments could come under pressure. Such a situation would also accentuate the risks of regional tensions and conflicts. This could produce unintended supply disruptions due to domestic or regional turmoil. On balance, this risk, although always present, might now be greater than before.

There is, however, another consideration which points to a similar conclusion. This time, it concerns OPEC policies. The erosion of real earnings from oil will produce frustration and a strong desire to compensate for 'lost income' in the past once the opportunity arises. Producer countries could thus, faced with a surge in demand fuelled in part by stockpiling, readily grab opportunities to maximize revenues per barrel. One might find in such a situation additional production ceilings at lower levels than up to now, a full exploitation of the opportunities of sales at spot prices, as well as other steps to exploit the occasion. A coordinated OPEC policy of moderation under such circumstances would be extremely difficult. Saudi Arabia alone could perhaps muster enough weight and spare capacity to prevent another price hike. However, we must doubt whether Riyadh would politically be able and willing to go it alone. There are at least some major uncertainties here not unrelated to future developments in the Middle East as a whole.

The probability of disruption will thus in my view be influenced negatively by market instability and falling prices. But what about Western European (and OECD) ability to absorb and manage shortfalls? One observation to be made here is psychological: falling oil prices will in all likelihood not encourage governments to take the energy security risks seriously enough. Thus, crisis preparations might slacken. Secondly, much will depend on the circumstances in which a crisis would occur. Both the first and the second oil shock were aggravated by rapidly rising demand for oil and energy while the non-event of 1980–1 in conjunction with the Gulf war had much to do with falling demand. This will be even more important in the future. As noted before, crisis management flexibility in terms of 'cutting the fat' will tend to decline in the future (although falling prices will slow the process). Refinery conversion, on the other hand, will increase flexibility, although again this trend could be slowed by the pressure on prices.

In a situation of supply shortages under conditions of potential upward demand pressure, price explosions could probably be contained only by a sophisticated use of stockpiles and/or recourse to OPEC spare capacity. This capacity might not be available for political reasons. A full exploitation of all elements of crisis management flexibility in both producer and consumer countries would probably be required to contain the impact of a major shortfall under such circumstances. This presupposes cooperation between the two groups. Whether such cooperation will be possible depends on a number of political preconditions. One of them might well be the degree of cooperation and dialogue *before* a crisis.

Overall, then, one certainly cannot be sanguine about the implications of growing market fragility for Western European oil security. The risks will change. They will be concentrated in periods of demand surges. But their dimensions will be as critical as before while government precautions against dangers of supply disturbances might slacken. Moreover, Western Europe will have to think about new energy security problems, in particular with respect to natural gas. This, too, will require some attention by governments. A third political challenge posed by the Western European energy security outlook concerns the issue of market volatility itself. The international oil market has never functioned under conditions of competition and fragmented control. The peculiarities of this market make it unlikely that it would then function well. Oil is too important a commodity to be left to market forces alone.

Notes

1. Cf., for example, Organisation for Economic Cooperation and Development (OECD), *Energy Prospects to 1985* (Paris, OECD, 1974).
2. International Energy Agency (IEA), *World Energy Outlook* (Paris, OECD, 1982).
3. *BP Statistical Review of World Energy 1983* (London, BP, 1984).

4. *OECD Economic Observer*, March 1982; IEA, op. cit., p. 64.
5. Esso, *Energy in Europe, Looking Forward to the Year 2000* (December 1982).
6. For a more detailed analysis about risks and vulnerability with regard to coal, natural gas and uranium see my *Energy, Minerals, and Security* (London, Macmillan, 1984).
7. Commission of the European Communities, *Communication for the European Council on Communities Natural Gas Supplies* (Brussels, 15 October 1982 COM (82)653 final).
8. IEA, op. cit., p. 184.
9. *Petroleum Economist*, June 1984, p. 217.
10. Ibid.

Chapter 3

The oil market and Western European energy security

Fereidun Fesharaki and Wendy Schultz

Despite a significant decline in oil demand, the Western European economies are and will remain a major market for oil. Indeed, Western Europe accounted for just under 40 per cent of world oil imports in 1984 and was the largest single market for internationally traded petroleum. In the same year, the Western European economies paid around $85 bn. for their oil imports.

As the uncertainty in the international oil market continues, the Western European nations remain heavily dependent on OPEC oil, particularly on oil from the Persian Gulf and North Africa. This heavy dependence, together with political uncertainties on the future of Europe's major oil suppliers, raises the question of energy security and policy choices open to the governments in the region. This issue is particularly significant at present as the current oil glut and the changing perceptions of future demand and supply for oil are leading many nations into a false sense of security.

The purpose of this chapter is to examine the following issues: the state of the oil market and the role of key OPEC nations; Western European oil dependency; and major policy options and vulnerabilities facing Western Europe. In a sense this chapter attempts to provide a glimpse into future scenarios of oil availability in the next ten to twenty years, while at the same time focusing on issues facing the Western European nations.

OPEC and the world oil market

The 1973–4 and 1979–80 oil price increases have been dubbed as the two 'oil shocks'. The 1980–3 decline in demand for oil certainly merits the title of the third oil shock. Non-Communist consumption of oil declined from 52.6 million barrels per day (mb/d) in 1979 to an estimated 45.7 mb/d in 1984. OPEC production, which peaked in

1977 at 31 mb/d, is expected to decline to less than 18 mb/d in 1984 and non-OPEC supplies exceeded OPEC supplies in the non-Communist world for the first time in 1982.[1]

The decline in demand was due to three main factors. First, the world recession; second, structural decline in demand for oil; and third, oil company de-stocking of inventories. None of these factors are likely to continue indefinitely. The world economy is already beginning to recover and oil company de-stocking has almost bottomed out. As far as the structural changes in demand are concerned, some irreversible changes in energy/GDP ratios have taken place. However, many analysts feel that the conservation cycle has run its course—at least with the present level of prices. The question of what has led to the decline in demand is still not fully answered, and we may well not find out about it until the end of the decade. We are still not sure whether the 'structural' portion of demand decline (as compared to the recessionary portion of demand decline), has been caused by the 1973–4 oil price shock or by the 1979–80 oil price shock. If, according to the World Bank, the long-term price elasticity of demand takes fifteen to twenty years to work itself through[2] (through the change of energy-using capital stock), then the full impacts of the first oil shock will be seen by the late 1980s and the full impact of the second oil shock will be observed between 1995 and 2000. This does not mean that demand will continue to drop, since the 'income effect' through growth in Gross Domestic Product, could neutralize the price effect.

For the medium-term outlook we have drawn on a number of studies made available in 1984 (see Table 3.1). These data provide a glimpse into the future energy mix through the eyes of leading forecasters. The forecasts expect a steady growth of 2 per cent per year in primary energy demand, but the distribution of energy mix is of particular interest. In the non-Communist world oil's share in supplying growing energy demand is expected to decline from 47 per cent to nearly 38 per cent in the year 2000. For the world as a whole a similar decline is projected. Gas and coal are expected to basically maintain their share, while large growth in hydro and nuclear power is foreseen. Other more exotic energy sources are not expected to have much of an impact.

The main message of the forecasts is that, though oil's share in supplying world energy is expected to fall, the drop is not great. By

Table 3.1 Mean of major projections of world energy demand* (million barrels per day of oil equivalent)

	1982	1990	1995	2000	Annual growth rates, % 1982–1990	1990–2000
Free world primary energy demand	92.3	112.7	127.7	137.2	2.5	2.0
Total world primary energy demand	136.7	170.1	190.6	218.0	2.8	2.5
Free world energy consumption by fuel type (%)						
Oil	47.0	43.2	40.8	37.9	1.4	0.7
Gas	18.8	18.8	19.0	18.4	2.6	1.7
Coal	22.0	22.1	25.1	24.3	2.6	3.0
Nuclear	4.2	6.5	7.6	8.6	8.2	4.9
Hydro and others	8.0	9.4	7.5	10.8	4.6	3.4
Total	100.0	100.0	100.0	100.0		
Total world energy consumption by fuel type (%)						
Oil	41.2	38.0	35.2	33.4	1.7	1.2
Gas	19.2	19.5	20.2	19.9	3.0	2.7
Coal	29.9	28.8	29.6	29.9	2.3	2.9
Nuclear	3.2	5.7	6.4	7.4	10.4	5.2
Hydro and others	6.5	8.0	8.6	9.4	5.4	4.2
Total	100.0	100.0	100.0	100.0		

*Based on data and projections by CONOCO, US DOE, Pace, IEA, OPEC, PEL, SRI, World Bank, and *BP Statistical Review of World Energy, 1982*.

the year 2000 oil still remains by far the world's most important source of energy. While the percentage of oil used will decline, the absolute volume will grow from 64 mb/d in 1982 to 83 mb/d in the year 2000—a growth rate of 1.7 per cent per annum between 1982 and 1990 and 1.2 per cent in the 1990s.

Demand and supply for OPEC oil

Demand outlook

The world oil demand has fallen from 63 mb/d in 1979 to around 57 mb/d in 1984. The 6 mb/d drop in demand for oil is accompanied by a 13 mb/d drop in OPEC oil output in the same period. In the non-Communist world demand fell from nearly 50 mb/d in 1979 to nearly 46 mb/d in 1984.

There are many forecasts of where oil demand will go in the future. Such forecasts are arrived at by utilizing large world oil models or by expert guesses. Except for a few who see a decline in demand for oil, most (including this author) believe the demand for oil will grow in the future. Any reasonable GNP forecast and energy/GDP ratio points to a growth in demand of between 1 to 2 per cent a year. Based on these assumptions we have constructed three scenarios of demand for oil through the year 2000. The demand scenarios and assumptions are shown in Tables 3.2 and 3.3. It can be seen that demand for oil in the non-Communist world will be 47.7–51.1 mb/d in 1990 and 52.7–62.3 mb/d in the year 2000. In all three scenarios, demand for OPEC oil will not exceed 23.6 mb/d in 1990—well below OPEC production capacity. Thus there will be little prospect of a surge in the price of oil this decade, under any scenario. For the year 2000, in the low case scenario and base case scenario, there will still be enough spare capacity to resist pressure for a big jump in the price of oil, but real price increase may start in early 1990s, under the base case. In the high case scenario, OPEC demand of 32.3 mb/d will have certainly led to real price increases in the 1990s, but the real price level could be in the range of 1979–80 price

Table 3.2 Scenario: free world medium- to long-term oil outlook (mb/d)

	1983*	Low case		Base case		High case	
		1990	2000	1990	2000	1990	2000
World oil demand	44.9	47.7	52.7	49.4	55.9	51.1	62.3
Oil supplies							
Non-OPEC†	24.3	25.5	26.5‡	25.5	26.5‡	25.5	26.5
CPE exports	1.6	0.5	0.5	0.8	0.5	0.8	0.5
Required OPEC oil output	17.0	20.8	24.5	22.1	26.2	23.6	32.3
OPEC NGL	1.0	0.9	1.2	1.0	2.7	1.2	3.0
Inventory change	(0.4)	—	—	—	—	—	—

* Actual.
† Includes processing gains and synthetics.
‡ Mean of expected output of 25.5 to 27.5 mb/d.

Table 3.3 Non-Communist oil demand: percentage growth rate

Low case		Base case		High case	
1983–1990	1990–2000	1983–1990	1990–2000	1983–1990	1990–2000
1%	1%	1.5%	1.25%	2%	2%

of oil—since demand on OPEC oil will only reach the 1979–80 range in the high case scenario.

Supply outlook: production strategies

Having discussed the demand side, let us consider the supply prospects. It is often pointed out that it is the large excess OPEC production capacity that is holding down OPEC prices. This is only partially true. OPEC excess capacity is not only a function of physical capacity but also the political will to produce. Clearly at production rates of say below 20 mb/d the issue of willingness to produce is not an important issue. However, as production rises, production ceilings will be reimposed. A relevant question to ask is what the OPEC production capacity is likely to be in the future. Will it fall, rise or remain stable?

OPEC nations' production policies are determined by the interplay of three major factors: world demand for oil; physical ceilings due to technical reservoir production capabilities; and policy ceilings on production. At any point in time, one of these three factors weighs heavier in determining production levels, but it is misleading to suggest that the OPEC nations respond only to oil demand or say, only to domestic, economic and political pressures in deciding production levels. Western analysts, basically used to studying private firms, often make the mistake of equating sovereign OPEC nations to private firms and try to 'model' their responses to market changes. This approach is not very helpful. Let us examine each of these three factors briefly.

First, oil demand and market responses have different significances for oil exporters depending on the status of the market. If the market is tight or in balance, OPEC will watch the spot market developments to adjust their prices upwards. If the market is in a state of glut with spot prices significantly below the official prices, they will strongly resist downward adjustments, trying to maintain normal prices while they let inflation erode the real price. As the current oil market softness is expected to last for a number of years, OPEC nations' most obvious option would be to curtail production. Those who can afford it, such as the Arab Gulf states (minus Iraq) will have to bear the brunt of cutbacks. So, the issue of response to a soft market is likely to come mostly through curtailment of production rather than price reductions. The major problem in this situation is the distribution of losses in production. Contrary to the conventional wisdom in the West that OPEC production and price policies aim to slow down the longer-term switch out of oil, we contend that this consideration plays only a minor role in OPEC's decision-making process. The decision-making process is simply not geared to the luxury of long-term concerns, because short-term and medium-term considerations are dominant. For instance, responses to the market are influenced greatly by the short-term impact on the world economy, but not so much by whether solar energy or coal liquefaction will displace petroleum. In short, response to oil demand is problematic for those in immediate need of foreign exchange, though the remedy is seen as curtailing production with the bulk of curtailment falling on the richer nations.

Second, there are clearly physical limits to output which place a ceiling on oil production. Outside of the Gulf states, all OPEC members have relatively small reserves and none could sustain production at 2 mb/d for more than a few years. If the market softness continues for a prolonged period of time, the physical ceilings will not be a determining factor until the 1990s: however, if demand picks up, the physical pressures to curtail production might emerge by late 1980s.

Third, policy ceilings—that is, curtailment of production for political and economic reasons—are mainly applicable to the Gulf states which have the largest reserves and potential production capability (perhaps with the exception of Qatar). Saudi Arabia and Kuwait have, for some time, placed official and unofficial ceilings on their production. The United Arab Emirates maintains 'allowables' (ceilings), and post-revolutionary Iran is restricted to a maximum ceiling of 4 mb/d, and has in the past preferred an output level of around 3 mb/d. Iraq has a physical ceiling of 3–4 mb/d and might be prepared to go up to that level if there is sufficient demand.[3]

Physical limitations

In looking at OPEC supply potential, a good starting point is the physical limitation the exporters face. There are many studies which have addressed this question. Here we would only briefly refer to two studies by Rand Corporation[4] and the US Department of Energy.[5] Nehring's well-known study estimates ultimate conventional world oil resources by region. This study identifies the Persian Gulf as not only having the largest proven oil reserves, but also the largest potential from future discoveries and enhanced recovery. Table 3.4,

Table 3.4 Estimated ultimate conventional world oil resources by region (billion barrels)

Region	Known	Potential	Total†
North America	179.8	100–200	280–380
South America	68.4	52–92	120–160
Western Europe	24.6	25–45	50–70
Eastern Europe/Soviet Union	102.4	63–123	165–225
Africa	75.6	45–94	120–170
Middle East	509.9*	350–630	860–1,140
Asia/Oceania	50.8	54–104	105–155
Total	1,011.5	688–1,288	1,700–2,300

420–730 Enhanced recovery† ← 688–1,288 → 263–555 Future discoveries†

* Of this figure 503.3 billion barrels are located in OPEC Gulf.
† Totals do not add up due to rounding.
Source: Richard Nehring, *Giant Oil Fields and World Oil Reserves*, op. cit., p. 88. Date as of 1 January 1976.

adapted from Nehring, estimates that around 50 per cent of potential addition to world oil reserves is likely to come from the OPEC Gulf states. According to Nehring, enhanced recovery alone could add 250–400 bn. barrels to proven Middle East reserves—equal to around one-third of all potential additions to world reserves. However, the same study sees a much smaller addition through enhanced recovery for other OPEC nations. For instance, increased recovery for North African nations is estimated at 15–30 bn. barrels.

Another major study by the US Department of Energy is less optimistic about prospects for enhanced recovery in most OPEC countries. This study is less comprehensive than the Rand study in that it does not consider either new discoveries or tertiary recovery. However, it is much more detailed and focused–providing for the first time information which has been regarded as confidential for decades. For instance, DOE figures indicate secondary recovery of only 87 bn. for the Persian Gulf. Since tertiary recovery is not expected to increase the recovery factor to much above 50–60 per cent of oil in-place, the Nehring estimates would be around twice as much as DOE estimates, even if we add 50 per cent to DOE estimates for tertiary recovery. Such a wide difference is, however, not that unusual when it comes to predicting future additions to oil reserves. If we had to choose among the two estimates, we would tend to lean towards the DOE study for the following reasons. First, the DOE study is undertaken and managed by people who are known to be competent to the author, and who have had access to Petro-consultants' data files, which were evidently not available to Nehring. Second, the latest information was obtained by DOE through the United States Geological Survey and the oil companies. Third, in the case of Iran, their data is very close to confidential Iranian data, which were not generally made public. This in itself is one test that they may not be too far off the mark. Table 3.5 summarizes the broad conclusion of the study on the Persian Gulf. An important characteristic of these data is the relatively high rates of primary recovery and low rates of secondary recovery. Iran, for instance, has the second highest volume of oil-in-place. However, recovery factors are very low, giving a lower remaining recoverable oil supply to Iran than Kuwait. (Indeed in another study on Libya and Algeria, DOE data indicates very low rates of secondary recovery of no more than 2.5 per cent for each country.) Outside of Iran and the neutral zone, the total of primary and secondary recovery factors are not that far off from the United States' average of 35–40 per cent. The Gulf nations have a remaining reserve of around 439 bn. barrels. This is around 90 bn. barrels, or 25 per cent, more than their current proven reserves. From the point of view of oil importers, it is important to note that for policy planning purposes the additions to OPEC oil through secondary recovery, though substantial, are less than previously assumed and that these indications should clearly be

Table 3.5 Reserves and resources of crude oil from known fields of the Persian Gulf Area as of 1 January 1979

Pertinent data	Saudi Arabia	Kuwait	Iran	Iraq	United Arab Emirates	Divided Neutral Zone	Qatar	Total OPEC Gulf	Oman	Bahrain	Total Gulf
Original oil in place, mb	550,583	258,446	437,063	127,240	151,640	75,954	26,472	1,600,925	26,256	2,718	1,656,372
Proved ultimate recovery, mb	179,658	95,874	72,100	47,193	37,972	12,914	8,390	454,101	4,550	1,005	459,656
Cumulative production, mb	33,718	18,986	28,227	13,630	5,884	2,968	2,836	105,249	1,273	651	108,173
Proved oil reserves, mb	145,940	76,888	43,873	33,563	32,088	9,946	5,554	347,852	3,277	354	351,483
Indicated additional recovery, mb	31,321	14,135	20,400	3,391	12,745	3,262	2,100	87,354	0	0	87,354
Total ultimate recovery, mb	210,979	110,009	92,500	50,584	50,717	16,176	10,490	541,455	4,550	1,005	547,010
Remaining recoverable oil, mb	177,261	91,023	64,273	36,954	44,833	13,208	7,654	435,206	3,227	354	438,837
Primary recovery factor (%)	32.6	37.1	16.5	37.1	25.0	17.0	31.7	28.4	17.3	37.0	27.7
Secondary recovery factor (%)	5.7	5.5	4.7	2.7	8.4	4.3	7.9	5.5	0	0	5.3
Ultimate recovery factor (%)	38.3	42.6	21.2	39.8	33.4	21.3	39.6	33.9	17.3	37.0	33.0

Source: US Department of Energy, *Middle East: Crude Oil Potential From Known Deposits*, op. cit., p. 9.

taken into account for assessing longer-term petroleum supplies. Tertiary recovery might indeed add 150 to 200 bn. barrels to recoverable oil in the Gulf, but it is as yet too early to say much about it. It is possible that like secondary recovery, tertiary recovery will not increase the recovery factor by the expected 10–20 per cent, and in fact, we will not know for sure how much more oil can be recovered until we have tried it.

For policy planning purposes, the gross volumes are less important than expected production profiles. That is to say, how long can a country produce oil at a specific level and what is the most likely level an individual country will choose? This approach uses logistic functions and decline curves to show the range of possibilities.

In the following we will briefly discuss the reserve and production profile of each country of the OPEC Gulf. In each case it is assumed that indicated additional reserves can be brought on stream in time for the scenarios under consideration. It is also assumed that sufficient investments worth tens of billions of dollars can be made in that time span. Figures 3.1 to 3.6 consider alternative production schedules for six OPEC Gulf nations. For Saudi Arabia a production rate of 12 mb/d would mean a decline by 1998, while production of 8 mb/d could be maintained until 2019. In Kuwait, a 6 mb/d could

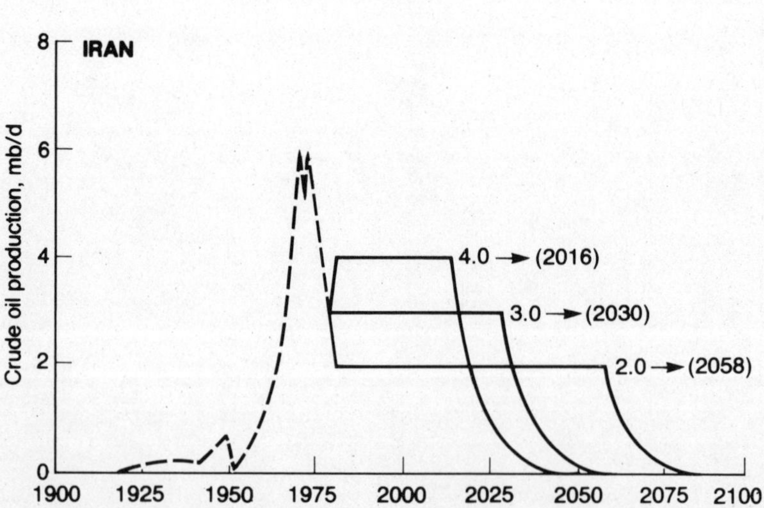

Figure 1 Historical and projected production profiles based upon the ultimate recovery from known oil deposits in Iran.

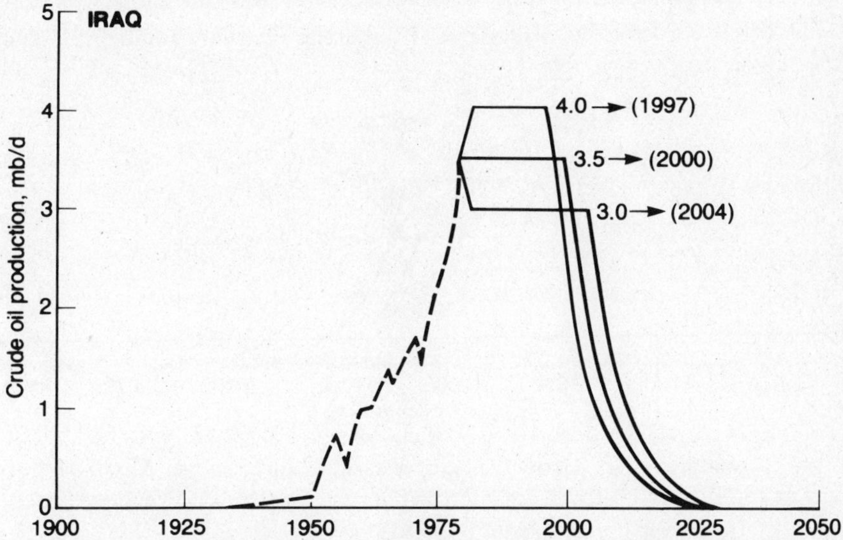

Figure 2 Historical and projected production profiles based upon the ultimate recovery from known oil deposits in Iraq. *Source*: Modified from *Middle East—Crude Oil Potential from Known Deposits* (Washington, DC, US Department of Energy, 1981).

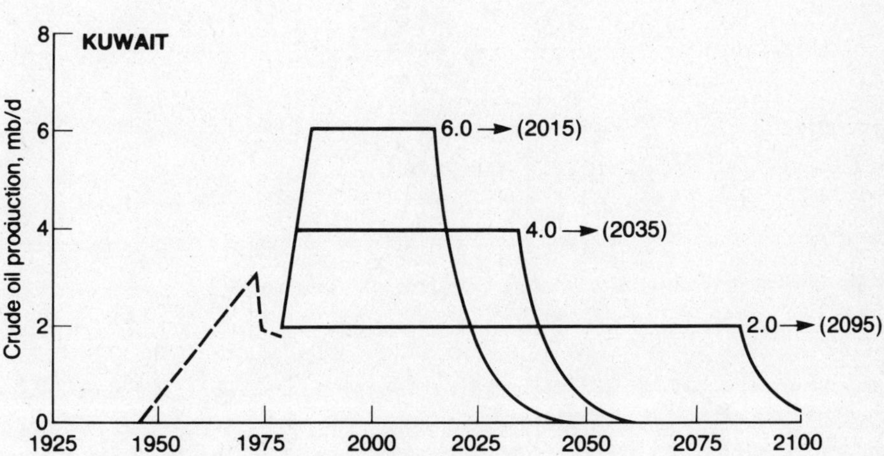

Figure 3 Historical and projected production profiles based upon the ultimate recovery from known oil deposits in Kuwait.

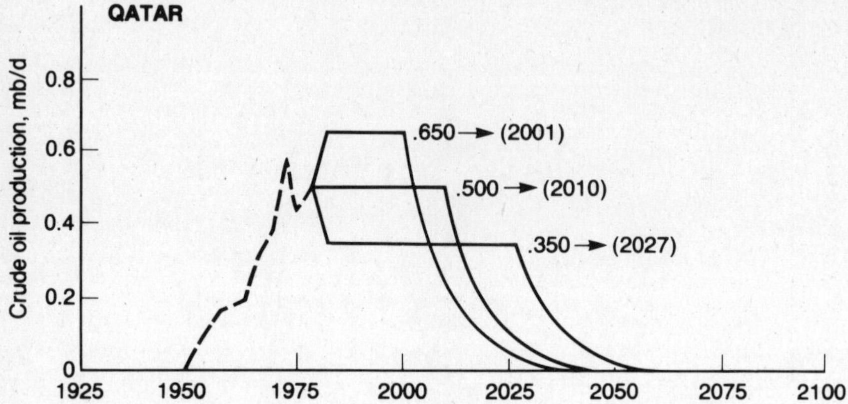

Figure 4 Historical and projected production profiles based upon the ultimate recovery from known oil deposits in Qatar. *Source*: Modified from *Middle East—Crude Oil Potential from Known Deposits* (Washington, DC, US Department of Energy, 1981).

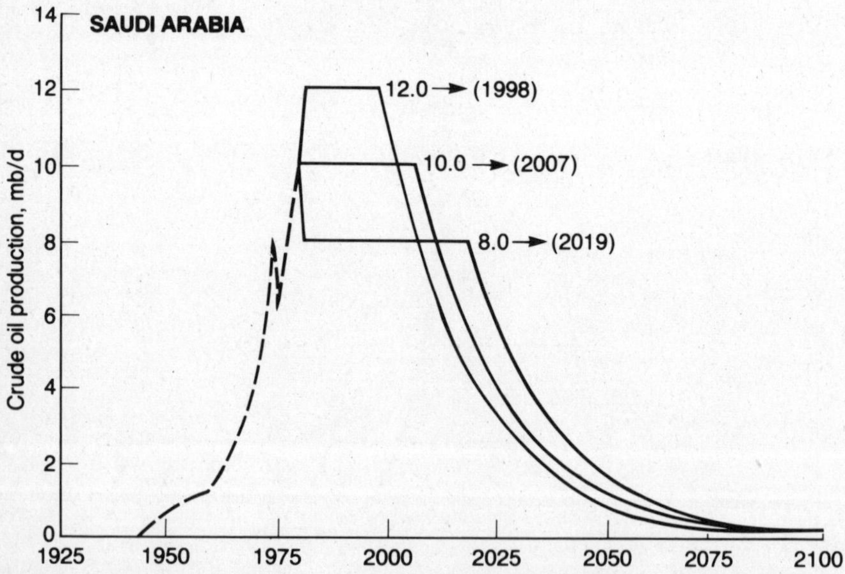

Figure 5 Historical and projected production profiles based upon the ultimate recovery from known oil deposits in Saudi Arabia.

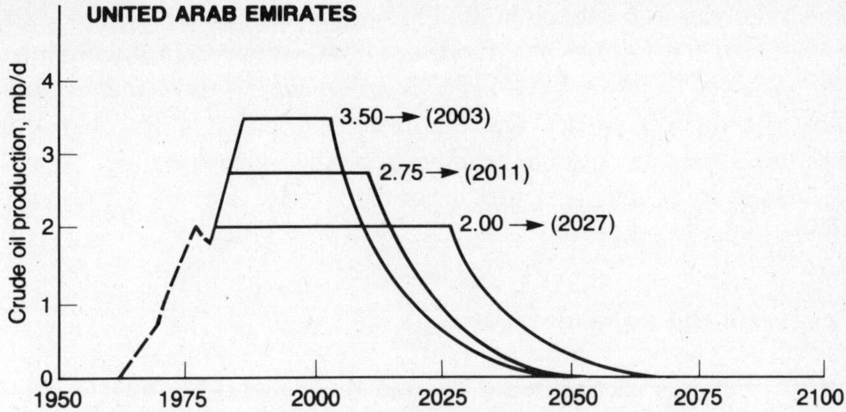

Figure 6 Historical and projected production profiles based upon the ultimate recovery from known oil deposits in United Arab Emirates. *Source*: Modified from *Middle East–Crude Oil Potential from Known Deposits* (Washington, DC, US Department of Energy, 1981).

only continue until 2015, when decline would begin, while a 2 mb/d could last until 2095. For Iran, a production level of 4 mb/d could be maintained until 2016, but at a 3 mb/d level, the decline would set in around 2030. In Iraq, a 4 mb/d output could not be sustained until the end of this century and output will begin to decline by 1997, however, if a 3 mb/d production is targeted, it could be maintained until 2004. In the United Arab Emirates, the higher output level of 3.5 mb/d could not be maintained beyond 2003, but a 2 mb/d level could last until 2027.

The above discussion is, of course, based on known deposits and potential additions from secondary recovery. Possible additions to reserves from tertiary recovery and new discoveries are not included in these analyses. Clearly, higher reserves could affect the shape of the curves significantly. However, many experts both within and outside OPEC do not expect major new discoveries of the size experienced so far in this century. As to the impact of tertiary recovery on proven reserves, there is little agreement among experts, simply because we do not know much about it. Before we close this brief section on reserves, a few words must be said about Iraq reserves. Since the mid-1970s, there has been a great deal of speculation that the size of Iraqi reserves may be much larger than what is reported in the trade press–possibly near the Saudi levels.

This speculation has been fuelled by vague statements from the Iraqi leadership that Baghdad is 'floating on oil'. We remain unconvinced that massive deposits have been in existence but essentially hidden from the outside world. There is little doubt that more oil will be discovered in the Arabian–Iranian geological province, but there is no reason to believe that new Iraqi discoveries are going to be substantially larger than the others.

Economic and political factors

In the previous section we discussed the general boundaries of oil production possibilities in the key OPEC nations of the Persian Gulf. Here we will briefly supplement our discussion of the physical limits with some of the general economic and political factors which are superimposed on the physical constraints of these nations.

Iran

Iran's peak production of 6.5 mb/d in the mid-1970s would have declined to 3–4 mb/d after 1985, even if the Shah had survived the Iranian revolution.[6] The decline in production could have been slowed down if the massive gas injection secondary recovery program that the Shah had started had not been suspended by the revolution. As things stand in May 1985, Iran's production cannot go much above 3–3.5 mb/d in 1985 (in line with the Iranian government's declared oil production ceiling preference), and by 1990 this production could be maintained if parts of the gas injection programme are resumed. As shown in the earlier section, Iran's known deposits are capable of maintaining an output level of 3–4 mb/d well into the twenty-first century. Given the structure and the size of the Iranian economy, Iranian planners estimate a need for foreign exchange of around $15 bn. a year (in 1981 prices) for 1982–5, and possibly $20 bn. in 1985–90, at a minimum. This necessitates export levels of between 1.3 to 1.7 mb/d in 1985 and 1990 respectively—a range well within the capabilities of Iran. As domestic demand is expected to be around 600,000 b/d in 1985, and about 800,000 b/d in 1990, and as Iran would need to build up her now-depleted foreign exchange reserves, we estimate a production level of around 3.5 mb/d during the 1980s. For the year 2000, rising domestic oil consumption and population growth may lead to a production level of 4 mb/d. We feel Iran's oil production policy in the mid-range is to a great extent independent of who rules the country.

Iraq

Iraq's resource base and secondary recovery prospects do not allow the country to sustain a production capacity of much beyond 3.5 mb/d in the 1980s. Although no clear information exists as to the minimum foreign exchange requirements of Iraq, past budget and expenditure data suggest that this level of output generates sufficient cash for Iraq's needs. These levels would not allow grandiose economic programmes simultaneous with massive military buildup. For the year 2000, Iraq's production is expected to decline to 2.5 mb/d as recoverable reserves fall. Like Iran, Iraq's production profile is partly independent of the regime which rules the country.

Kuwait

Kuwait's large recoverable reserves permit this small nation the greatest flexibility in the Gulf. Kuwait can produce up to 6 mb/d of oil and continue at this level until 2015; however, it is highly unlikely that Kuwait's output will ever exceed 1.5–2.0 mb/d. Kuwait's production capacity still remains at around 3 mb/d, but production has consistently been below that level since the mid-1970s. Kuwait's acute awareness that oil remains its only major export product, and that its limited production (indigenous population is only 40 per cent of the total population of around 1.2 million) and land base will not permit economic diversifications inside the country, has pushed Kuwait to be the most conservation-minded OPEC nation. In the past, Kuwait has imposed production ceilings of 2.0 and 1.5 mb/d. The last formal ceiling announced was 1.25 mb/d. In 1984, Kuwait's output fell below 1 mb/d, not because of new ceilings, but because of demand declines resulting from recession, conservation and oil company stock drawdowns. It is expected that Kuwait may be producing below its preferred ceiling for a year or two, which might result in some drawdown of foreign reserves. To compensate for such drawdowns, Kuwait may prefer to produce 1.5 mb/d by the mid-1980s and maintain that level through the year 2000. The higher level of output compared to the present ceiling was arrived at by considering the unusual growth in domestic consumption of oil products in Kuwait. In 1981, Kuwait had the second highest energy consumption per capita in the world; at 60 barrels oil equivalent per person, Kuwaiti consumption was second only to the United States. Although consumption growth is likely to slow down, we feel that a ceiling increase to 1.5 mb/d is likely. The Kuwaiti refining capacity would require around 700,000 b/d of crude for processing. This level can, in fact, be regarded as a floor to the likely Kuwaiti production. The preferred level of 1.5 mb/d could well be maintained into the early twenty-second century before decline sets in—and this is based on known deposits only.

Kuwait's tremendous flexibility also stems from her foreign exchange accumulations. By the end of 1984, Kuwait had amassed around $50 bn. and return on income was more than $6 bn. in 1981.[7] Indeed, Kuwait could cease production and survive for many years. If production is continued at the expected levels, some time early in the twenty-first century Kuwait's income from investments may well be equal to her oil revenues.

Unlike Iran and Iraq which suffer from domestic instability, Kuwait is relatively stable. Although there are some dissenting voices which led to a period of suspension of the parliament, the massive influx of oil revenues has ensured the nation's stability—despite the existence of around 750,000 guest workers.

Insofar as international politics are concerned, Kuwait remains tied to the West, but with a decidedly independent line. It has proved to be the most future-conscious Gulf nation, with a generally clear idea of where it is going. Kuwait's acute awareness of its vulnerability to foreign attacks or invasions has made it the only nation in the Gulf Cooperation Council to seek cordial relations with the Soviet Union, although it is unlikely ever to leave the pro-Western camp. Like Iran and Iraq, Kuwaiti oil policy may again remain independent of the leadership of the country. No matter who rules Kuwait, the projected production levels are likely to prevail.

Qatar

Qatar's small recoverable reserves mean the small nation's output is likely to decline slightly during the remainder of this century from their current figure of about 500,000 b/d. Qatar does not pursue a particular oil policy, and depends on the general Saudi or Kuwaiti line of policy in OPEC. Like Kuwait, Qatar is relatively stable internally, but has the same vulnerabilities to foreign interference. Output in Qatar is not dependent on the leadership of that country.

United Arab Emirates

UAE is a federation of seven small emirates. The largest portion of reserves and production comes from Abu Dhabi (1980 production: 1.35 mb/d), a smaller portion from Dubai (1980 production: 0.35 mb/d), and a very small fraction from Sharjah (1980 production: 0.01 mb/d).

UAE's recoverable reserves from known deposits permit a production of 2 mb/d before decline sets in in 2027. UAE's preferred level of output is around 1.8 mb/d, although no official government policy has been declared. We expect the level of 1.8 mb/d to continue through 1990. By the year 2000, UAE's preferred level of output might rise slightly to 2 or 2.5 mb/d, to accommodate rising domestic demand, and the subsidies to non-oil producing members of the federation.

Abu Dhabi is the principal financier of this loose federation—put together at the urging of the British after their 1971 withdrawal from the Gulf. The income gap between the haves and have-nots has been a source of tension in the federation. If the federation is to survive, Abu Dhabi will have to continue to act as bankers for the poorer states. Internal opposition in UAE is minimal within individual nations, although some degree of hostility is apparent between the emirates. If the federation does not survive, Abu Dhabi's oil production may fall by 0.5 mb/d due to smaller budgetary needs. Thus, the range of 0.5 mb/d is, in our view, the maximum volume of change which may take place in UAE's output.

UAE's policy and foreign policy is very much a carbon copy of that of Saudi Arabia. On a number of occasions UAE has been the only supporter of Saudi policies in OPEC.

Saudi Arabia

The large resource base and recoverable resources of Saudi Arabia put the Kingdom in the position of dominating the world oil scene well into the next century. Saudi Arabia's self-imposed production ceiling of 8.5 mb/d during the 1970s (with some deviation above the ceiling after the Iranian revolution and the Iraq–Iran war), was reduced to 7.0 mb/d in April 1982 in the face of declining demand for oil. By the end of 1984 Saudi output had fallen to below 4 mb/d.

Saudi Arabia's policy of moderation on the OPEC front has been the subject of much debate in the past decade. Saudi moderation, defined as high output and lower prices (lower than other OPEC nations would have liked) has been typically (and inaccurately) attributed by many analysts to: (a) the Saudi 'enlightened' self-interest—i.e. not wanting to price their large oil reserves out of the market and cause massive shifts into non-oil energy sources; (b) Saudi interest in economic stability in the OECD nations, since recession and economic problems weaken the industrial world, which is the natural political ally of the Kingdom; (c) Saudi interest in the economic well-being of the Western world, in which all of their estimated foreign assets of $140 bn. are invested, as any economic damage would automatically depreciate the value of their own foreign investments; (d) the Saudi 'special relationship' with the United States—that Saudi moderation on the oil front is rewarded by a pledge to protect the Kingdom against foreign military actions, and by a United States' promise to help resolve the Palestinian issue with Israel; and (e) the Saudi budgetary needs, which some analysts claim require high levels of production that might not find markets at higher prices.

The arguments outlined above are a mixture of fact, half-facts, and fantasy, forwarded by different interest groups in the West, or coming from Western analysts who are simply unaware of the decision-making process in the Middle East. Our own subjective interpretations of the Saudi policies are that, first, the Saudis are not interested in destroying the economies of the industrial world or their own assets, but neither are the other OPEC nations. The Saudi interest in seeing the world economy in a reasonable state is a factor in their own oil policy. Second, the Saudi budgetary needs are often misunderstood and misinterpreted by various groups. The confusion between domestic currency and foreign exchange, resulting from conversion of Saudi riyals to US dollars, has created major over-estimates of Saudi needs. Indeed, the Saudis have made it clear since 1981 that they could continue their present pace of development with production of 4–5 mb/d. The Saudis also enjoy great flexibility with their massive foreign exchange holdings, which we estimate resulted in a return on investment of $12 bn. in 1981. They can also slow down their development projects with negligible impact on the economy (many of the capital investments are located in industrial enclaves with no real linkage to the domestic economy); and, of course, they can draw on their foreign reserves. Though short-term budgetary pressures may appear every now and then, we feel it pointless to try to base Saudi oil production projections in revenue needs alone. Annual budget and five-year development plans are really only guidelines, and it is rare that actual numbers ever match. Indeed, we believe that short-term difficulties, such as a budget deficit, and a balance-of-payments deficit in Saudi Arabia, will result in rationalization of expenditures and will not have the longer-term impact of pushing production to higher levels. Finally, we feel that the United States–Saudi connection provides the most relevant explanation of the Saudi policies: the expected defence of the nation from foreign attack, and defusing the Arab–Israeli time bomb.

While an Iran-type revolution in Saudi Arabia is unlikely, discontent, particularly among the educated middle classes, is apparent. Even within the Royal Family there are many who believe that overt cooperation with the United States is neither in the best interests of the country nor the regime. They argue that the United States has not been willing or able to deliver a solution to the Palestinian problem, and that overt friendship with Israel's main ally destabilizes the regime and makes it look like a puppet of the United States. None of these groups wish a break with the United States, but many believe in following a more independent Arab line. To show such independence they would favour production levels below the historical rates.

Taking into consideration all arguments—political, economic and technical—it is the author's view that Saudi preferred production level is unlikely to exceed 6 mb/d during this decade. By the year 2000, rising domestic consumption could push Saudi production to around 8 mb/d.

A scenario of OPEC oil production

Table 3.6 shows a profile of OPEC production capacity through the year 2010. For 1984 and 1990, the data represents preferred levels of output or acceptable levels of output in the present oil market. It takes into account national priorities, attitudes, political/social preferences as well as technical limits. For 2000 and 2010, the data mainly relies on technical production limits, given known resources, enhanced recovery, and probability of additional new discoveries. Clearly each country can produce more, but it would then suffer faster decline rates. The choice of production levels is clearly arbitrary, but we believe the figures provide a feasible scenario.[8]

Table 3.6 shows that in 1984, available supplies for almost all non-Gulf OPEC members are close to maximum capacity. In the same year, available supplies are likely to exceed demand by 6 mb/d, leaving a large enough cushion to exert downward pressure on demand. Available capacity in 1990 will be around 28.5 mb/d or at least 5 mb/d higher than what may be demanded from OPEC. By the year 2000, in the low and base case scenarios, there will be still sufficient excess capacity to stem strong upward price pressures; but in the high case, there will be price increases. As we approach the year 2000 price pressures will become stronger.

In short, there will be sufficient OPEC capacity to produce oil demanded from OPEC without a major price increase well until late 1990s. OPEC oil availability is improved by the major slowdown in OPEC domestic oil demand. Demand growth in OPEC is estimated at 5 per cent per annum until the end of this decade, almost one-half of the historical rates, dropping to 3.5 per cent per annum in the 1990s, and 3 per cent in the period 2000–10. The decline in the growth of OPEC consumption is expected through higher domestic prices, gas substitution for oil, and a general move away from oil in power generation.

Coping with the short term

The 1970s was certainly the decade of OPEC. OPEC members were continuously looking forward to higher prices and higher output. When the price of oil was raised by 30¢ a barrel in 1971, there were

Table 3.6 Preferred/expected levels of oil production in OPEC countries (mb/d)

	1984	1990	2000*	2010*
Iran	3.2	3.5	4.0	3.5
Iraq	1.2	4.0	3.5	4.5
Kuwait	1.2	1.5	2.0	2.0
Saudi Arabia	6.5	6.5	8.0	8.0
UAE	2.0	2.5	3.0	3.5
Qatar	0.5	0.5	0.5	0.2
Neutral zone	0.5	0.5	0.4	—
Total Gulf	15.1	19.0	21.4	21.7
Indonesia	1.6	1.8	1.3	0.5†
Libya	1.7	1.7	2.5	2.5
Ecuador	0.3	0.5	0.5	0.3†
Venezuela	2.2	2.5	2.5	2.5
Gabon	0.2	0.2	0.2	0.2†
Algeria	0.8	0.8	0.8	0.8†
Nigeria	2.1	2.0	1.3	0.5†
Total	24.0	28.5	30.5	29.0
Domestic consumption‡	3.3	4.4	6.2	8.3
Exportable Oil	20.7	24.1	24.3	20.7

* Technical limits taking into account production profiles and decline rates, except for Kuwait which has the greatest flexibility. Assume capital investments will be forthcoming as required.

† Possible oil importer by 2010.

‡ Based on F. Fesharaki, L. Totto and T. Johnson, 'OPEC Domestic Demand Revisited', *Petroleum Economist*, London, September 1984.

cries that the world economy would fall into the recession and oil demand would decline. It didn't! At every step of price increases (whether it was OPEC induced or OPEC just following the market), it seemed that the market could absorb the increase and demand would still rise. Clearly, OPEC did go too far, but no one could scientifically show where the optimal price for oil might have been. It simply had to be a trial and error procedure.

The ever rising nominal (if not real) level of revenues led to expectation of continuous expansion of expenditures. Thus, when the production decline came, all OPEC nations faced difficulties in adjusting their expenditures downwards. Such difficulties were more psychological or organizational than real. After all, the budget growth was so fast that one could always find enough fat to reduce the budget without serious shortcomings. Many rich, powerful men who benefited from the high expenditure levels through commissions and kickbacks had to become content with not increasing the size of their bank accounts! Altogether, this was a rude awakening for the relatively sloppy, inefficient expenditure mechanism in many oil exporting nations.

The response to the lower output was pressure for higher production. But of course, higher production could not create demand and the markets weakened. At the conclusion of the OPEC meeting in January 1983, Saudi Oil Minister Yamani stated his belief that the oil price would not hold. This led to the massive stockdraw of 4–5 mb/d in the first quarter of 1983 which brought about the London agreement that reduced the price of oil to $29 pb for Arab Light.

It is ironic that the reduction in the price of oil became the glue to OPEC unity. With production levels of 15–16 mb/d, OPEC nations feared they could lose significant levels of revenue through further drops in oil prices. They chose to stick together and stop further losses in revenues. If one is forced to try to make a general case for the short-term OPEC behaviour, one may make the following generalizations:

- Fear is the glue to unity.
- Fear is most effective at production levels of around 16–17 mb/d.
- Fear works if there is the possibility of a drop in the price of oil.
- At output levels of 20–21 mb/d for OPEC oil, production is high enough to satisfy the members, though most would prefer 24–25 mb/d of output.
- Between 16–17 to 20–21 mb/d, price structure is *inherently unstable*. Cheating on prices and production will go on and members will compete for sharing out the incremental demand.
- Effect of fear wears off without a continuous threat to a price cut or decline in output to 16–17 mb/d range.
- Maintaining discipline requires a policeman, namely Saudi Arabia, which does not currently seem willing to take that responsibility.

If one applies the above general rules to our previous discussion, one can explain the weakness of the market in the summer of 1984.

In fact, by July 1984, every OPEC country, with the exception of Kuwait, was violating OPEC rules either by overproducing or by giving discounts on crude and products. The massive slide of $2–3 pb in spot prices was halted through a production cutback and major effort by non-OPEC producers and the oil industry to support the marker price.

Essentially, the fear of price collapse was not so present in OPEC minds in the late summer of 1984 and there was much hope that higher seasonal demand in the fourth quarter would see OPEC through the year end. OPEC's problems are, however, more fundamental. Since demand for OPEC oil is unlikely to exceed the critical point of 20–21 mb/d until the late 1980s, this means that price instability will continue over the next couple of years. Weak second and third quarter demand will exert downward pressure on the price of oil. Meanwhile, the necessary discipline is simply not there and there is a possibility of a price decline in nominal term.

How much of a decline? It will be surprising if the decline is more than $3–4 pb. Why should the price not fall further? Prices will not fall more because it is clear that lower prices will not resolve OPEC's difficulties in the short to medium term. OPEC will just end up collecting less revenues. Why $25 pb? Many OPEC experts are beginning to think that at such a level new substitution away from oil will cease. Whether this is true or not, such is the perception of a number of senior experts. At the same time, non-OPEC exporters and the oil industry will do all it can to stop a big drop in the price of oil.

It is often argued that oil prices could fall to the market clearing price, whatever that may be. This argument is often advanced by those who have little appreciation for the difference between a private firm and a sovereign state. In the words of Paul Frankel, 'If OPEC is a cartel, it is in a class of its own'. The usual rules applied to the private sector do not always give us good enough tools to measure OPEC's viability as a cartel. Also, implicit among most arguments about OPEC collapse is the idea that OPEC oil ministers are not intelligent enough to foresee the results of their own actions. OPEC ministers and their experts are, in fact, very capable people. They have a good feel for the oil market and they often buy the best talents too. It is patronizing to assume they are less able to work out their options than some analysts in the consuming world.

OPEC's major challenges in the years ahead

For OPEC nations the future holds new challenges raised by the change in the structure of the oil market and irreversible declines in certain uses of oil. The most important of these new challenges are summarized below.

Need for continued pro-rationing

Two years ago Allrio Parra of Venezuela said that OPEC is moving toward a more 'structured' market system.[9] This was meant to imply that pro-rationing or production control may become an integral part of OPEC policies for a number of years. Events in the recent past have confirmed the validity of this assertion. It now seems inevitable that production control needs to be in force until demand for OPEC oil recovers to above 20–21 mb/d. As stated before, it may well take a few years for that to come about.

In short, OPEC will have to act far more like a cartel than it has in the past. Inevitably, some members will break the rules, but as long as it is not excessive, a little cheating can be tolerated. The countries that may cheat will do so only if they are sure they can get away with it, for if other nations reciprocate, the benefits from cheating will diminish and all OPEC members will lose. In effect, the cheating member will behave like a non-OPEC member, but only at a level that will not substantially hurt the organization and its goals. It does not require elaborate economic analysis to show that lowering the prices will not increase demand in the short term, and, if every nation reciprocates, everyone will have lower revenues and not much higher output. OPEC nations have clearly shown that they understand such short-term problems of price and demand. Ironically, OPEC's willingness and ability to administer the market may turn out to benefit the consumers and oil companies as well as the producers—as long as such administration does not result in large increases in price. If OPEC breaks up and the prices are left to market forces alone, the upward and downward swings in price will inevitably become a lot more volatile than they would be with OPEC administration. In fact, OPEC's role will be that of a 'shock absorber' moderating the peaks and troughs of price movements. There is little doubt that a five-year or ten-year period of low oil prices would stimulate demand for OPEC oil, push out substitute fuels such as coal or renewables, kill the synthetic fuel industry, and price out of the market new, high-cost non-OPEC oil. But one should not lose sight of the fact that oil is finite, and that to expect it to remain at low prices if new exploration is halted, demand rises, and existing proven reserves are depleted, is naïve. Indeed, in another study, we have shown that even in a perfectly competitive market, long-term oil prices might not fall much lower than the 1982 level of prices.[10]

OPEC's new challenges in the 1980s will put to the test the fundamental strength of the organization and its membership's resilience. The new price regime of the oil market is likely to be inherently more unstable than in the past. Most OPEC members, and the Saudis in particular, are interested in a smooth, balanced market and are likely to do their utmost to achieve this objective. In the event of OPEC's failure to administer the market successfully, all parties are likely to lose. OPEC's task of controlling the market could be made much harder by the short-sighted view of some governments or oil companies that elect to agitate for the breakup of the organization. The non-OPEC oil exporters should keep in mind that without OPEC, they themselves will be forced to come together and carry the burden of oil market administration.

Impact of downstream investments

It is useful at this stage to explore the impact of OPEC's downstream ventures on the future of the organization. First, the move into downstream activities signals an important change for OPEC's position in the oil market. Unlike the past, when OPEC remained in control of production and expected the oil companies to refine, transport, and market the members' crude oil, the OPEC nations will now have to 'get their hands dirty' to reach the end users. Such new involvement will bring the product exporters much experience in dealing with the consumers, and at the same time will make them much more aware of the market realities and pressures that play on the oil companies. This new turn of events is likely to push some OPEC members to become more market-orientated. Since not all OPEC members will be product exporters—some will remain wholly in crude oil exports—friction between those that export only crude and those that export both crude and products is likely to emerge.

However, because the large crude producers with the largest reserve potentials are going to add refining capacity, it is logical to foresee that OPEC downstream ventures will make the whole organization more market-orientated.

Second, the administration of the market may become difficult in the short term, as new product exports will have to capture a space for themselves in the competitive product trade in the consuming areas. However, in the medium term, OPEC's downstream ventures will make price and production administration easier. In effect, we are likely to see the national oil companies of some OPEC nations emerging as vertically-integrated major oil companies—and vertical integration by its nature implies that tighter control of the market can be exercised.

Third, control of crude and refined petroleum products, and some control of transportation, open up the possibility of 'packaging' a number of export items together. Clearly, a prerequisite for OPEC's capability to 'package' is that excess refining capacity world-wide be substantially reduced, if not totally eliminated, through scrapping. Packaging is, of course, easier in times of tight markets, and more difficult in times of glut.

Fourth, OPEC members' entry into the product market is expected to lead to a more consistent and market-orientated pricing policy. OPEC product exporters will have to compete in the market with their own crude and with other products. This will necessitate taking into account product prices and refinery margins in relation to crude prices. In short, product netbacks must have a logical relationship with crude prices. In the *medium term*, the chances of a repetition of the illogical price relationships between crude oil and products that prevails when product prices often fall below crude prices and cause major refining losses, are likely to be substantially reduced when OPEC itself is a partner in product marketing. At the same time, the uneconomic price differentials for crude in 1979–80, when African Light crudes were priced $9 pb above the Saudi Light, cannot survive in a market where around 20 per cent of OPEC's exports are in product form (see Table 3.7). The crude price differentials are likely to be constrained to levels consistent with technical refining costs. Any deviations from such technical differences would result in major losses to one or a number of OPEC exporters.

Table 3.7 Projected oil product exports, OPEC and the GCC nations, 1987–8 ('000 b/d)

Exporter	Refined product production	Domestic demand	Gross product exports
Bahrain	225	16	209
Kuwait	690	95	595
Oman	45	29	20
Qatar	55	16	41
Saudi Arabia	1,655	614	1,040
UAE	160	125	35
GCC Total	2,830	895	1,940
Iran	975*	740	0–235*
Iraq	385*	430	0*
Other Gulf total	1,360	1,170	0–235
Algeria	390	130	260
Ecuador	75	127	–
Gabon	15	44	–
Indonesia	810	550	260
Libya	315	146	170
Nigeria	230	221	9
Venezuela	1,220	420	800
Non-Gulf OPEC total	3,055	1,638	1,490
Total, Gulf & OPEC	7,245	3,703	3,430–665
OPEC Total	6,975	3,660	3,200–435

* Uncertainty resulting from the Iran–Iraq war.
Source: F. Fesharaki and D. Isaack, *OPEC and the World Refining Crisis*, The Economist Intelligence Unit, London, 1984.

Finally, OPEC's downstream operations will enhance its political control of crude and its international influence in the 1990s. As OPEC moves toward vertical integration through partial control of other stages of oil processing and delivery, its flexibility in imposing its will on others increases. For instance, two important factors that led to the defeat of the Arab oil embargo of 1973—namely, the international oil companies' power to switch crudes to different refineries and to use their control of the tanker transportation network for deliveries–will be less of an influence by the 1990s. Oil product exports carried by OPEC product carriers, and crude exports carried by an OPEC-owned and chartered tanker fleet, will make switching crudes and products more difficult. Clearly, the vulnerability of the oil-importing nations rises with OPEC's move into the downstream market, but this may be counterbalanced by an increasing alignment of the interests of the OPEC and OECD nations. Because the Saudis remain the largest product and crude exporters, their goodwill and their political stability will have an important effect on where the market will go.

The argument that downstream investments will make OPEC members more market-conscious is not inconsistent with the argument that OPEC must prepare itself to administer the market. As in socialist theories of mixed economies, the market signals can be used to improve the administration of the market and to avoid major misallocations. This is not the optimum free-market approach, but the oil market never really was a free market. Moreover, the current structure of the industry may require market administration to avoid massive financial losses to both oil exporters and oil importers.

The role of non-OPEC producers

There are many non-OPEC oil producers and a number of important exporters. As shown in Tables 3.8 and 3.9 in 1983, non-OPEC nations produced 26.7 mb/d of petroleum. This output rose by 1 mb/d in 1984.

Important non-OPEC exporters such as the United Kingdom, Mexico, Soviet Union, Egypt, China and Malaysia produce around 17.5 mb/d of oil but export just over 4.5 mb/d or a quarter of their output. These six countries together have exports amounting to nearly 20 per cent of all petroleum trade in the world. In many respects, this key group of countries is even more of a heterogenous group than the OPEC membership itself. However, they have been able to use the market better than OPEC—because they are not bound by OPEC rules. Generally speaking, both within and without OPEC, the oil exporters are intent on getting the best price they can for their crude. However, the more radical are the governments, the more they have acted in a market-orientated fashion! Short-term

Table 3.8 Non-OPEC supply (mb/d)

	1983	1984	Comments
North Sea	3.0	3.3	New 1983 fields reach maximums
Mexico	2.7	2.8	Exports still restricted to 1.5 mb/d
Canada	1.4	1.4	Stable
United States	8.7	8.7	Stable
NGL*	3.4	3.6	Gathering systems to increase
Net CPE exports	1.6	1.6	High exports continue
Other†	5.9	6.3	Increases from West Africa, Egypt, Brazil, India and Malaysia
Total	26.7	27.7	

* OPEC and non-OPEC excluding centrally planned economies.
† Includes 0.8 mb/d processing gain and 0.15–0.20 mb/d synthetics.
Source: Gulf Oil Corporation, July 1984, Houston, Texas.

and long-term considerations may create divisions at times between the exporters, but it is foolhardy to believe that they are suicidal! Non-OPEC producers used the tight market of 1979–80 to sell at substantially above OPEC prices and then undercut OPEC prices during the weak market. Even within OPEC certain members have used *de facto* discounts to increase output at the expense of a core group. It is crucial, however, to note that the non-OPEC producers can only take advantage of the OPEC prices if OPEC is there to shoulder the burden of market pressures. Every time there is a real danger of OPEC collapse, non-OPEC producers fall into line.

Since March 1983, Mexico has behaved as a *de facto* OPEC member with a self-imposed export ceiling. Indeed, Mexico has behaved better than most OPEC members. The United Kingdom has held the line on price but is legally unable to restrict production over a long period of time. The crisis of July 1984 was a classic case of serious danger facing the oil prices. Not only did non-OPEC exporters, with the exception of Soviet Union, resolve to protect the prices but also the major oil companies showed their first serious effort at bolstering the prices.

Within the non-OPEC exporters, Mexico is the most important. Taking into account enhanced recovery and additional discoveries,

Table 3.9 Non-OPEC supply—'other' (mb/d)

Country	1983	Expected 1984	Change in production
Angola	175	195	20
Argentina	480	485	5
Australia	410	450	40
Brazil	325	450	125
Brunei	155	160	5
Cameroon	115	120	5
Colombia	150	170	20
Egypt	720	750	30
India	390	435	45
Malaysia	370	440	70
Oman	380	400	20
Peru	170	175	5
Syria	180	180	—
Trinidad	160	155	(5)
Tunisia	110	110	—
Zaïre	25	30	5
Other countries (24)	625	665	40
Grand total crude	4,940	5,370	430
Processing gain & synthetics	950	950	
Total 'other'	5,890	6,320	

Source: Gulf Oil Corporation, July 1984, Houston, Texas.

Mexico's production profile demonstrates that Mexico could produce at 2.5 mb/d for 100 years. A production level of 4.5 mb/d could be sustained until 2030. It is reasonable to assume that Mexican output will rise to about 4.5 mb/d by the mid 1990s—if the market recovery warrants new investments—but exports may stay in the range of 2–2.5 mb/d, given the strong growth in domestic oil demand. In short, Mexico will fall in the major OPEC league of oil exporters.

The United States is another country which deserves a mention. Output of 8.7 mb/d is expected to decline slowly, making the United States more dependent on imports. Although the United States is an oil importer, the size of its domestic demand and the variety of crudes processed have given it an important influence in price determination world-wide. Despite declining output levels, such a role for the United States will continue for many decades to come.

Western European energy dependence

Western European nations are major consumers of oil. Oil consumption in the region rose significantly in the 1950s and 1960s but slowed down considerably in the 1970s. Indeed, during 1980–3, demand for oil declined in most European countries.

Table 3.10 provides a broad picture of oil and energy use in thirteen European countries. It can be seen that the oil consumption declined from 13.1 mb/d in 1974, to 11 mb/d in 1983. Portugal and Spain increased their oil consumption slightly, while others, particularly United Kingdom and West Germany, showed large declines. As it was stated earlier on, we do not know as yet how much of the decline resulted from the economic recession or structural changes in energy efficiency. Many analysts, however, believe that there is not much more scope for energy efficiency in many European countries, and that higher rates of economic growth will once again increase oil demand in Europe.

The contribution of oil to energy consumption has also declined in Europe: from 58 per cent in 1974 to 48 per cent in 1983. Though this decline is important to note, nevertheless, one should not lose sight of the fact that in 1983 approximately one-half of energy use in Western Europe was supplied by one type of energy: oil. Though it is not possible to say with certainty whether the share of oil in energy demand will fall much further in Europe, a number of observations may be made. First, within OECD nations, Europe is still the most energy efficient region, an area which has pushed conservation and alternate energy use to near maximum limit, thus leading many observers to conclude that further movement away from oil will be slow. Second, even assuming the average expected OECD rate, it is unlikely that Europe's demand for oil declines below one-third of its energy use by the year 2000.

In considering net oil imports as a percentage of oil consumption in major European economies, Table 3.10 shows the dependence declining from nearly 100 per cent in 1974 to less than 75 per cent in 1983. This decline, though impressive, still leaves Europe highly

Table 3.10 Western European oil consumption (thousands of barrels per day)

Western Europe	Oil consumption*		Oil as % of energy consumption		Net imports as % of oil demand†	
	1974	1983	1974	1983	1974	1983
Austria	210	205	45	42	78	76
Belgium and Luxembourg	560	435	56	48	99‡	95‡
Denmark	320	215	88	66	108	75
France	2,460	1,880	66	49	104	87
West Germany	2,760	2,320	52	44	97	87
Italy	2,015	1,810	74	65	98	96
Netherlands	725	620	50	43	79	63
Norway	160	155	27	23	79	–
Portugal	135	195	73	78	94	85
Spain	820	990	67	61	94	75
Sweden	540	405	63	51	103	91
Switzerland	270	255	58	46	99	92
United Kingdom	2,135	1,530	49	37	108	–
Total Western Europe (13 countries)	13,110	11,015	58	48	99	73

* Includes naphtha, LPG, gasoline, kerosene, refining fuel, and bunkers.
† Where demand is over 100 per cent, it indicates inventory build-up.
‡ Percentage slightly in error due to lack of Luxembourg data.
Source: *BP Statistical Review of World Energy, 1984*, London 1984, and *Oil and Energy Trends*, **4**, No. 12, and **10**, No. 5, London, 1979 and 1985.

dependent on oil imports. It should be noted, of course, that almost all the decline in dependence is explicable through North Sea production and sales to other European countries.

A measure of the cost of oil imports is shown in Table 3.11. Between 1974 and 1983, the value of net oil imports rose from $52.4 bn. to $122.2 bn. (without North Sea production, the cost would have risen by an additional 25 per cent). As the table shows, the oil import bill has been the single largest component of the balance of trade deficit in many European countries. For instance, in 1983, Belgium, Denmark, France, Italy and Spain would have had a trade surplus if they had had no oil imports. Except for the United Kingdom, at least 44 per cent of the balance of trade deficit of all European countries was caused by oil imports.

Table 3.12 shows the dependence of Western European nations in

Table 3.11 Oil and trade balance in Western Europe (US$m.)

Western Europe	Value of gross oil imports		Percentage of trade balance deficit due to oil imports*	
	1974	1983	1974	1983
Austria	765	1,790	41	44
Belgium	2,337	7,970	160	236
Denmark	1,466	2,530	68	809
France	10,380	20,390	160	313
West Germany	11,292	25,310	(36)†	(134)†
Italy	7,971	21,300	75	231
Netherlands	1,962	13,550	166	(91)†
Norway	–	–	–	–
Portugal	467	2,010	21	52
Spain	3,016	10,200	36	104
Sweden	2,444	5,620	(96)†	(285)†
Switzerland	1,336	3,000	52	84
United Kingdom	8,975	8,560	58	(123)‡
Total Western Europe (13 countries)	52,411	122,230	–	369

* Where it exceeds 100, it indicates that the trade balance would have been in surplus without oil imports.

† Oil importers and trade surplus country; the figure shown indicates the percentage increase possible in the country's trade surplus were it not paying for oil imports.

‡ Oil exporter and trade deficit country; the figure shown indicates the percentage increase in the country's trade deficit were it not exporting oil.

Source: *Oil and Energy Trends Statistical Review 1982*, and *Oil and Energy Trends*, **10**, No. 5, London, 1982 and 1985.

oil imports from the Persian Gulf and North African OPEC members. While dependence on the Persian Gulf was generally reduced (mainly as a result of North Sea output and the emergence of non-OPEC oil trade), still the dependence remains extremely high and dangerous. Except for West Germany, Sweden and Switzerland, Europe depends on the Gulf for one-quarter to two-thirds of its oil imports.

Algerian and Libyan crudes compete directly with Nigerian and North Sea oils. As the latter are of similar quality but usually are less expensive, they bite into North Africa's market share. In the last two years, however, shifting price differentials have closed the price gap between these competitors. As a result, Western European dependence on North African oil actually increased slightly from

Table 3.12 Western European dependence on oil imports from the Persian Gulf and North Africa,* 1974 and 1983 (percentage of total gross imports)

Western Europe	Persian Gulf		North Africa		Total	
	1974	1983	1974	1983	1974	1983
Austria	54	24	12	25	66	49
Belgium	86	30	8	12	94	42
Denmark	42	30	3	1	45	31
France	78	38	11	13	89	51
West Germany	43	19	21	22	64	41
Italy	86	43	20	17	100	60
Netherlands	100	37	–	15	100	52
Norway	80	–	6	–	86	–
Portugal	58	62	–	5	58	67
Spain	88	49	12	12	100	61
Sweden	18	5	1	0	19	5
Switzerland	28	17	11	24	39	41
United Kingdom	72	26	8	2	80	28

* OPEC member countries only.
Source: *World Oil Trade*, December 1981, and *Oil and Energy Trends*, **9**, No. 12, London, 1984.

1974 to 1983. In sum, by 1983 Austria, France, Italy, the Netherlands, Portugal and Spain were dependent on the Persian Gulf and North Africa for at least 50 per cent of their total oil imports.

Future prospects: options and vulnerabilities for Europe

The world oil market turmoil has affected the perceptions of many policy planners. Many people feel that the current glut has enhanced the energy security of European nations. They expect the glut to continue, with demand falling and OPEC as well as non-OPEC supplies rising. This may well harm the longer-term policy options and growth prospects of the European nations.

The analysis in this chapter points to a number of factors. First, Western Europe's dependence on oil imports, particularly from the Persian Gulf and North Africa, is likely to remain dangerously high throughout this century. Any embargoes or disruption of supply from the key suppliers of oil to Europe is likely to have a devastating impact on the European economies. Second, the production profiles and reserves analyses point to the fact that the Persian Gulf nations will continue to remain the major actors in the world oil market well through the early part of the twenty-first century, if not longer. North African nations will remain important until 1990, but by the turn of the century, their lower reserve base would imply less crude available from these nations and consequently more dependence on the Persian Gulf. Third, expectations of continuous additions to non-OPEC supplies are unrealistic. Some additions will no doubt be forthcoming but in much smaller volumes. North Sea output can at best be expected to remain the same at current rates through 1990 and then decline. Mexican output increases will be matched by massive increases in domestic consumption, which is expected to rise to 2 mb/d by 1990, leaving net exports of only 500,000 b/d above present export levels. Consequently, the oil market's major actors will remain the Persian Gulf nations. Fourth, Western European refinery capacity utilization remains low. Such excess capacity will come under even greater pressure when OPEC export refineries begin to enter the market. The natural market for the export refineries of the Gulf is Western Europe. Many billions of dollars of surplus capacity have been scrapped or shut down in Europe. More such closures are inevitable. At the same time, European imports of refined products from the Gulf will make European energy security even more fragile than it is today.

The options open to European economies are limited. At the same time, it should be noted that vulnerabilities of the region to oil supply disruptions are only *marginally* reduced by the decline in oil demand. Long-term energy policy should become divorced from short-term glut or shortage considerations and directed more toward policies which maintain the momentum of energy efficiency and supply diversification. However, the momentum of energy efficiency and diversification cannot be left to private economic forces. Specific and clear-cut government policies are called for. The European nations cannot rely only on the International Energy Agency's policy recommendations, or short-term political expediency within their own borders. More use of coal and natural gas should be encouraged through government policy. Refinery scrapping and rationalization must be undertaken in spite of union pressures or nationalistic sentiments to keep some inefficient and outmoded refineries open. Diverse sources of supplies of oil and gas should be sought. Unfortunately, such policies must emanate from coordinated European action, a type of coordinated action which has had little precedence in modern Western European history.

More importantly, the European nations should realize that their future energy security is still highly dependent on actions of the Persian Gulf oil producers. Bilateral and multilateral relations both on the economic and political front should be encouraged and expanded between Western Europe and the Gulf producers. Furthermore, it goes without saying, that Western Europe's energy security is greatly dependent on political stability in the Persian Gulf at least for the next twenty years. Ever present threats of domestic upheavals and, more importantly, the possibility of foreign aggression in the Gulf are direct threats to European security. The European nations should closely monitor the political and economic development in the Gulf and use their influence, whenever possible, to contribute to regional security.

Notes

1. F. Fesharaki, '1982 Outlook for Oil', *Middle-East Economic Survey*, **25**, No. 30, 10 May, 1982.
2. World Bank, *World Development Report 1981*,Washington, DC.
3. F. Fesharaki, 'World Oil Availability: The Role of OPEC Policies', *Annual Review of Energy*, **7**, 1981.
4. Richard Nehring, *Giant Oil Fields and World Oil Resources*, Rand Corporation, California, 1978.
5. *Middle East Crude Oil Potentials from Known Deposits*, US Department of Energy, Washington, DC, 1981.
6. F. Fesharaki, *Revolution and Energy Policy in Iran*, The Economist Intelligence Unit, London, 1980.
7. *Petroleum Intelligence Weekly*, 29 March, 1982.
8. F. Fesharaki and S. Hoffman, 'Determinants of OPEC Supplies: Physical, Economic, and Political Factors', presented to the International Association of Energy Economists, November 1982.
9. A. A. Parra, 'OPEC Move May Lead to a Structured Market', *Petroleum Intelligence Weekly*, 12 April, 1982.
10. J. R. Roumasset, D. T. Issak and F. Fesharaki, 'Thinking About Oil Prices', *Energy Economics*, July 1983, pp. 164–70.

Chapter 4

The Soviet Union as energy supplier

Antony Scanlan

In the last ten years very fundamental changes have occurred in the world energy market. Twelve years ago, at Easter 1973, American oil import quotas which had had the effect of largely divorcing the largest oil-consuming nation from the world of oil trading, were dissolved overnight. The basic reason for this was the failure of fifteen years of protection aimed at stimulating relatively high-cost American indigenous oil production to a level necessary to maintain an effective level of self-sufficiency for the United States. Some increases in United States' oil imports had been allowed throughout the period and, as a result, between 1958 and 1970, United States imports rose from about 1.5 mb/d to 3 mb/d; but production then peaked and in the following two years imports doubled again. Quite simply, a stimulus to raise production without any equivalent stimulus to curb consumption could not continue. Are we now observing a similar phenomenon in respect of the Soviet Union?

During this same period from 1970 (when production in the United States peaked) Soviet oil production doubled making the Soviet Union the leading oil producer in the world. As a result of this ability to raise production, the Soviet Union and Comecon oil markets remained a self-sufficient system and were therefore only marginally related to oil trading in the rest of the world. But during this period consumption in Comecon continued to rise almost in parallel but unlike the Soviet Union the other Comecon nations do not have the ability to increase oil production. As a result, Soviet exports outside Comecon remained very broadly on the same level, contributing some 4–6 per cent to supply in the non-Communist world.

During the period since 1970 the rising demand for oil in Eastern Europe has twice caused a rise in non-Soviet imports of oil. The first episode occurred after the first world oil crisis in 1973, and the second was more noticeable, lasting from 1975 for the next four years. The major problem that this trade posed for a non-convertible currency area was the need to develop and sustain hard currency exports. Could this be done?

End of a period

An abrupt answer to this question appeared in 1979, when oil prices on the world market exploded for the second time and imports into Eastern Europe from non-CMEA sources fell away, and indeed have not yet recovered. But the change has not been confined to Eastern Europe. I would like to suggest that the world energy scene in the 1980s—and therefore the economic scene—has changed as significantly from the 1970s as the 1970s differed from the 1960s. This is as true of the Comecon bloc as it is for the rest of the world.

At the same time there is a strong persistent theme throughout all three decades. It applies as much to the Soviets as to the West. It is a simple fact that the days of cheap, easily accessible petroleum supply, which can double every decade in order to sustain the economy, are over. For the non-Communist world that period ended with the 1960s: for Comecon it ended with the 1970s.

In both cases the supply projections—surprising as it may sound to some ears—have been reliable if uninspiring: it is the *demand* projections and therefore the *economy* and *world trade* that have been forced to adapt. In theory, if adaptation is flexible and sensitive, there is no need for crisis. Expensive oil and other sources of energy are still available, though at a cost, and the use of these should encourage efficiency and lower energy consumption in every aspect of the economy. The proportion of each dollar spent on investment instead of consumption has to go up but the potential of energy efficiency should alleviate any slowdown in the economy. Why then, are both Comecon and OECD now obliged to face lower growth rates in the economy—say 2 per cent or 3 per cent per annum compared with twice or three times that rate which they previously enjoyed? Is there some common endemic oversight in both the major world economic systems? This question was put by one of our leading economists to his Soviet counterpart at a conference recently. It drew the cryptic comment 'Tell me, comrade, did *your* authorities listen to *your* geologists?'

One reason why 'doubling every decade' continued longer in the

Soviet Union is that the Soviet Union was a relatively late starter in major oil and gas production compared with the non-Communist world. At the turn of the century, Tsarist Russia produced more oil (largely from the Caspian Basin) than any other nation, but the early Five Year Plans in the 1920s and 1930s under Stalin favoured coal. When Stalin died in 1953, Soviet oil production had re-centred on the Ural–Volga fields developed during and since the Second World War, because they were less exposed to German advances into the Northern Caucasus; but Soviet production in total was only about 30 million tons per annum (mtpa), i.e. 600,000 barrels per day (b/d). By 1970 this had risen tenfold, still dominated by the Ural–Volga area—about the size of Iranian production at its peak. During the 1960s however, vast new deposits were discovered at the 'Third Baku' in the West Siberian plain. Virtually the whole of the subsequent development of All-Union production centred on this region ever since. Soviet production doubled since 1970 despite a progressive decline in Ural–Volga output since 1975. This is almost solely due to production developed in West Siberia, in particular the super-giant field at Samotlor which alone provides 3 mb/d or one quarter of Soviet oil, now again leading the world at 12.2 mb/d.

Natural gas has followed a very similar pattern, although, as I shall describe later, the reserve base is much greater than it is for oil. Production has risen tenfold since 1960. The earlier Caspian, Ukrainian and Ural–Volga fields have been superseded by immense discoveries in Central Asia, Orenburg Oblast in the extreme east of the European Soviet Union, and in the west Siberian plain up into the Ob Delta region, where the Urengoi and Yamburg fields are among the world's largest.

Shift to Siberia

The West Siberian plain now provides over half of All-Union production of both oil and gas. Its potential is still rising, especially in natural gas. All-Union production of gas is planned to rise by over one-third during the 1981–5 Five Year Plan, mainly as a consequence of development in this area. Soviet oil production growth, however, is close to its peak because of continuing decline in the 'First and Second Baku' regions, the Caspian and Ural–Volga. But 'Third Baku'

Siberian developments continue to provide enough to offset decline in the older areas. Minor production continues to be developed offshore from Sakhalin in the Far East, offshore from Lithuania, in the Ukraine, the eastern shore of the Caspian in Kazakhstan, in the extreme north-east European Komi and Pechora region and minor discoveries even as far afield as Kamchatka. But the success of the Five Year Plan depends almost totally on West Siberia. It is a modest plan for oil–about 20 million tons growth over the whole period whereas in the past decade the *annual* increment in the oil output was as large. By 1990, on the evidence available from Soviet sources, West Siberia will be providing two-thirds of both oil and gas production. In 1985 the amount of All-Union gas production will roughly equal the oil production in thermal value.

The scale of Soviet oil and gas production and that of West Siberia in particular can best be gauged by comparing the Soviet energy balance with that of the United States. Currently–1984 at present being the latest complete figures for total comparisons–the two nations' energy volumes are equally impressive: the United States produces about 31 mb/d of energy. This assesses hydro and nuclear power at the high factor, i.e. fossil fuel displacement into electric utilities for both production and consumption purposes; and equates all fossil fuels at oil equivalent (oe). United States energy consumption is about 36 mb/d, requiring oil imports of 5 mb/d, partly offset by coal exports. Soviet energy production is about 31 mb/d with consumption of 26 mb/d. Clearly, it is a major exporter of energy, of which 3 mb/d is oil, making it at present second only to Saudi Arabia among oil exporting countries. Its gas exports constitute the largest single international movement of gas in the world, 42 mtoe.

However, when Eastern Europe is added to the Comecon internal energy balance, apart from a net surplus of coal (in normal times for Poland) the Soviet net export surplus is considerably reduced. Eastern Europe has to import 80 per cent of its oil and over half of its gas, with scant prospects for increasing production. To summarize, the Soviet Union produces 12 mb/d of oil, consumes 9 mb/d and exports 3 mb/d. But 2 of those 3 mb/d are absorbed in other Communist countries–80 per cent in Eastern Europe and the rest in small amounts by North Korea, Cuba, Vietnam, Mongolia, etc. I do not include these, nor Yugoslavia nor Albania in the definition of Comecon in the rest of this paper–only the Soviet Union and 'the

Six'–Bulgaria, Czechoslovakia, the GDR, Hungary, Poland and Romania. Comecon, on this definition, still enjoys a surplus position–indeed energy is the main export stock of the Bloc. In total, Comecon energy production is 5 mb/d oil equivalent (oe) greater than the United States and consumption 3 mb/d lower.

Declining revenues

With 75 per cent of the Soviet population of 270 million located in the European Soviet Union and 75 per cent of each of the three main fossil fuels in the Asiatic Soviet Union, the need to develop an energy system of equal magnitude to that of the United States (and also to supply the 100 million people in Eastern Europe) and lacking the possibility of waterborne transport–there are no seaways in the Soviet Union and even the rivers run north-south–has necessitated a massive trunk pipeline network being developed over the past twenty years. It is the Soviet equivalent of the VLCC tanker fleet and just as essential. The main trunk systems with which we are most familiar are the twin Druzhba (Friendship) oil trunk export lines and the Bratsvo (Brotherhood) twin gas lines and the Soyuz (Union) gas lines which all principally export to Eastern Europe.

These lines are dwarfed by the internal grid for both oil and gas that has been developed and that links Central Asia to Leningrad, the Ob Delta to the Ukraine, and the Caucasus to Moscow. Its pivotal region is the area between Kuibyshev, Gorky and Chelyabinsk. Many of these lines are of 2,000/3,000 miles in length, the majority 40 inches or more in diameter and many routes have been looped. I estimate about 200,000 kilometres of major lines have been laid in the last twenty-five years. The much-debated Urengoi Export Line for Western Europe is only one of five major pipelines under construction in the current Five-Year Plan. It is now under way and about to speed up its recorded pipelaying rate of ten miles per day or more, despite the concurrent claims of the other major trunk lines, for example to Kursk, south-east of Moscow. All these lines appear to be on schedule for completion as planned.

In 1985, therefore, the West Siberian/Ob Delta complex is producing oil and gas equivalent eight times the hydrocarbon production of Alaska. How many people have noticed that Soviet oil production in the past two years has exceeded the total Middle East output of oil or that Soviet energy production now equals in total the highest annual OPEC production ever recorded (1977)?

There is, however, a big difference in the level of oil reserves behind such comparisons: Soviet oil reserves are a state secret. Various non-Soviet estimates place recoverable All-Union oil reserves at between 30 and 80 billion barrels. *Oil–Gas Journal* estimates oil reserves at between 60 and 70 billion barrels (10 per cent of total world reserves) and natural gas reserves at 40,000 billion cubic metres (bcm) (over 40 per cent of world reserves). Current levels of production are equivalent to twelve years of oil supply—slightly better than for the United States—and seventy years supply of gas. The Urengoi Export gas pipeline, at 35 bcm, will only represent 5 per cent of planned Soviet gas production by the end of the decade and substantial new reserves of gas are being added each year. The oil perspective is very different, the giant field of Samotlor, (15 billion barrels) is scheduled to decline by the year 1986. The total Soviet oil output implies depletion of nearly 5 billion barrels annually so that the equivalent of two more Samotlor fields will be needed simply to maintain oil reserves at current levels by 1990. If such new supergiants existed, we would undoubtedly have heard of them—Samotlor was first announced twenty years ago! Of course, the recent indication of a new group of fields in the less-explored north-eastern part of the West Siberian sedimentary basin, capable of 800,000 b/d three years ahead, is a sure indication that there is more to come in this vast basin—but supergiants are another matter. An ability to offset the Ural–Volga and Caspian decline, but no more than that, is implied by these new developments. Even more of them will be needed before 1990 as the Siberian fields maturing simultaneously with Samotlor will peak and then decline. Enhanced or tertiary recovery from existing fields is another possible way of maintaining production. However, it will take many years to develop beyond the 1 per cent of All-Union output that it represented as recently as 1980—and only certain options for EOR are open, i.e. are compatible with the ubiquitous practice of secondary recovery by water drive.

Further north and east there are still many potentially fertile basins awaiting development, and however remote or inhospitable the location, experience has shown that, if technically feasible, supergiants will be attractive even if they exist in the offshore Arctic or in Eastern Siberia or the far North-East. After many years, speculation about the Barents Sea, for example, is about to be ended with the current delivery of offshore rigs from yards in Finland, but it is now inevitable that the results will be seen in the next decade, rather than this one, in terms of commercial hydrocarbons. For the rest of this decade—as in many parts of the world—the system is limited by the time factor to existing development regions. Even with gas, the first indicators of planned targets for the 1986–90 Plan anticipate slower growth—about 10 per cent compared with over one-third in the current plan period. In 1986–90 it is intended that the major energy production increment will pass from gas to coal. This is not because the gas reserves are in question, nor is it simply because of production lead times—it is as much to do with the total energy structure intended and to the limits to effective large-scale substitution of gas for oil in the internal market. This is probably because Soviet policy is *not* to burn gas (except locally) in more and more power stations, a point often made by leading figures such as Minister of Power, P. Neporozhniy. The last oil-fired power station has been built in the Soviet Union and the future of electricity is based upon coal, hydro and nuclear power. Coal is intended to provide the main energy supply increase in the second half of this decade.

Coal

To put oil and gas in focus it is worth taking note of the geographical location of the coal fields. Three quarters of the coal reserves lie in Asia. Asiatic coal has taken over the major part of coal production. But growth has been insufficient to do more than offset the decline in the European Soviet Union with the Donetz Basin output now 10 per cent below peak. As a consequence All-Union production has remained static for nearly a decade. The low priority given to investment in deep mines in the past will take another decade to resolve, so hopes for the planned increase from 700 to 775 million coal tons by 1985 largely rested on 'soft' coals or lignite in the Asiatic Soviet Union and their use in power generation locally to be transmitted westwards by 'wire'. In the event, Soviet coal output in 1985 will show little increase over 1984—perhaps 725 million tons at best. (The increase in lignite in total output has contributed to a 25 per cent decline in the thermal value of the average ton mined in the past two decades.)

The decision to build power stations at the coalface has now committed the Soviet Power and Electricity Ministry to this option rather than to expanded railroads or slurry pipelines (even if these latter could be proved feasible over a long distance). The rail option, with nearly half of all rail freight traffic already dedicated to coal or to materials supporting infrastructure for mines, would severely overtax the rail system (which is still giving priority to the second Trans-Siberian railroad from Baykal to Amur–the BAM–rather than to modernization). Thus, even if the coal output targets are met, effective delivery depends upon long-distance electricity transmission. For this new high voltage direct current (DC) lines are being developed. They lose less power through voltage resistance over long distances than alternating current (AC) systems although they have to be converted into AC. The use of converters from DC to AC on such a large scale is a technical frontier for which equipment delays or failure to operate at full efficiency could be crucial. It is noticeable that dates for commissioning these lines have continuously receded twice during the current Five Year Plan. Neither the coal production targets for 1990 nor the All-Union Grid link-up is likely to be effective until the next decade when coal expansion expects to benefit from an investment priority it has not had for twenty-five years. But the planned increase for coal output indicated for the period 1986–90 was originally about three times the increase forecast for 1981–5. The 1985 target, although too high, was the same target originally set for 1980.

Another important aspect in the development of Soviet energy is that, because of the difficulties of 'transmitting' Asiatic coal, European parts of Comecon including the Soviet Union are scheduled for a most ambitious nuclear power construction programme. Output is planned to more than double within this Five Year Plan. While this is well within the technical capacity of Comecon, we are now told that 'Atommash', which had been planned to produce about eight reactors annually, will only add eight or nine new reactors in the Plan period, so that the official plan is clearly at risk of major slippage. If this occurs, then technical progress in 'coal-by-wire' will become even more important to compensate for delays in nuclear electricity.

In general terms, it may be a reasonable middle view of Soviet coal prospects that they will increase very little, perhaps 10 per cent, in *effective* energy contribution to the end of the decade. There are also likely to be delays to the nuclear programme. Thus the ability to prevent the continuing or expanded use of gas in power stations may be adversely affected. But just as the last oil-fired power station has been commissioned so it is intended to use gas to substitute for oil in all possible sectors other than power generation, both internally and for export—and that includes Eastern Europe. The extent to which oil to gas power generation switching occurs may be an indicator of delays in the coal and nuclear programmes.

The Soviet Far East

Before looking at Eastern Europe and westward prospects in general it is useful to focus on the eastward pattern—(essentially trade with Japan). It brings into focus another aspect of Soviet energy, namely, the *separate character of the Soviet Far East*. One has to travel 4,000 miles east of Moscow to reach the Pacific. Two-thirds of the journey is *eastward* of the major Siberian and Central Asian energy deposits. Two thousand miles East of Tyumen in the Yakutsk ASSR gas and coal discoveries have been made. They are, however, too far east of any large population centres to allow for any viable outlet except further eastward to Vladivostok and Japan. Offshore the smaller oil and gas deposits of Sakhalin offer better located but more limited prospects for these two markets.

Today Japan imports up to 8,000 barrels per day of Soviet oil and up to 1 million tons per annum of coking coal. The cities in the Pacific region such as Vladivostok constitute the only significant population in the 3 million square miles of the USSR east of Lake Baykal and are insufficient on their own to justify developments 1,500 miles north-west in Yakutia. The Japanese are therefore cast in the role of developer, since the mineral (and other) wealth of the Soviet Far East is much closer to Japan than it is to European Russia.

So far, Japan has accepted commitments to develop coal at Neryungi in Yakutia. It is scheduled to provide about 5 million tons of coking coal annually by rail to Japan via Nakhodka from 1985 to the end of the century: in the meantime, the coal is delivered from

Kuznetsk. The Japanese have not accepted a similar commitment for Yakutsk gas, preferring a trilateral deal involving the United States as equal importer. Even if the climate of acceptability to all partners were to re-emerge, the project would still take ten to fifteen years to develop. Each importer would then receive 7–8 bcm annually by LNG tanker from the port of Olga, north of Vladivostok. However, the project for Sakhalin LNG, from the mainland port of De Kastri, is due to commence by 1985 at 5 bcm annually. In this case the Japanese appear to have a readiness to proceed which is not affected by American attitudes.

Important as the infrastructure development may be for the Soviet Union, the quantities are tiny—about 1 per cent in the case of Japanese oil and coal demand and 9 per cent of Japanese gas requirements by 1985. The great bulk of Siberian resources are westward orientated; lying at the eastern edge of the populated area but still in the western half of the Soviet Union.

Eastern Europe's dependence

The vast majority of Soviet exports of energy goes either to Eastern or to Western Europe. This has not always been the case. In the 1960s a vigorous oil export programme included several Latin American, African and Asian countries. At the same time Eastern Europe had tripled its imports of energy—oil and gas—from the Soviet Union since 1968. By the early 1970s we were speculating on how the Soviet Union would choose between the developing world, hard currency markets (Western Europe and Japan) and CMEA requirements (Eastern Europe). It turned out that the developing world was first to go, although other forms of trade deal remained and some supplies of oil—not Soviet domestic oil—continued to come from these countries to meet Soviet third-party commitments elsewhere as rouble exports. Now, as gas supplies increase and oil exports face up to the problem that rising internal Soviet demand is threatening to match or exceed increases, the question is how the Soviet Union will make a choice between the two parts of Europe.

Speculation increased concerning the oil problem in Eastern Europe. In 1980 I wrote a paper which emphasized that something had to give within the Eastern European economy. Although, as usual, nobody could have predicted the explosive events in Poland that followed, they are by no means inconsistent with that view: nor has the process of adjustment in these economies to the new energy perspective yet run its course. How that course evolves depends to a great degree on the six Comecon countries themselves but to an unusually high degree it depends upon policy decisions in the Soviet Union. There is not space in this chapter to review all the energy aspects of each Eastern European nation. The key features in Eastern Europe are a unique combination of circumstances for a group of developed nations. Together they share an almost total dependence on imports for oil, an abnormally low percentage of oil in the energy balance and therefore a low ability to substitute other fuels for oil for certain key economic purposes. This is combined with membership of an economic union, the Comecon (or CMEA), whose most substantial export capability is, paradoxically, energy. To this must be added the fact that their use of energy per capita is one of the highest in the world, particularly in relation to their relative economic attainments. This explains their peculiar sensitivity to imported energy prices.

The three northern countries, Poland, Czechoslovakia and the GDR average a mere 20 per cent of oil in their primary energy mix. To find a comparison with the Polish energy balance of only 15 per cent of oil one would have to go to China. The southern countries average about one-third of oil in the energy balance, similar to the Soviet Union, but with Romanian production now in decline prospects are worsening. The great mainstay is solid fuels, mainly low grade outside of Poland and therefore only of use internally. The combined production of solid fuels in raw tons in a normal year is, at 700 million tons, equal to Soviet coal output. But the Soviet Union, with two-and-a-half times the population produces ten times as much gas and sixty times the combined oil output of Eastern Europe.

Since 1975 incremental Soviet gas supplies have exceeded increases in Soviet oil deliveries. As a consequence the terms of oil imports have gone from bad to worse. The rouble price of Soviet oil is linked to a five-year moving average of world prices. Before the world oil crisis in 1973 the CMEA importers had paid the Soviet Union a full arm's length price for their oil. Afterwards the five-year average system left them below world prices. Although in 1975 revisions doubled the 1973 price they still represented less than the full effect, certainly: but they were bad news for CMEA importers nevertheless. However, demand exceeded Soviet supplies. In the period from 1975 to 1979 Eastern European imports of non-CMEA oil quadrupled from 5 mta to over 20 mta. The second oil price explosion ended that at a stroke. In 1982 the Soviet Union imposed a 10 per cent cut overall, although some flexibility between national allocations in Eastern Europe was apparent. Total supplies were frozen at 80 mta (400 million tons spread over five years).

All these adverse movements in volume and price have dealt a severe check to economic growth. They coincided with Romanian oil production peaking and declining and were followed by the rescheduling of international loans by East European nations.

The answer to the higher real cost of oil imports must lie not in contraction but in an expansion of export trade, either with the Soviet Union or elsewhere. In the first of the energy papers produced by the British Institute Joint Energy Policy Programme, in the middle of 1982, Jonathan Stern drew attention to the likelihood of a rapidly diminishing gap between the price of Soviet oil delivered to Eastern Europe and the nominal equivalent in the non-Communist world price of crude oil. Since then Soviet prices have risen about 20 per cent. Current events in the world market add considerable speculation as to the pace at which this nominal gap may disappear or even invert. It has to be emphasized, however, that such nominal (i.e. arithmetical) comparisons of price do not in themselves represent the way in which the Soviet Union assesses the relative value of trade with Eastern Europe or Western Europe or indeed, how Eastern Europe assesses the relative value of Soviet or Western trade. But it is significant that the world parity price is already effective for quantities above the reduced Soviet oil supply volume available. What market economist could possibly object? Many countries in the developing world, equally caught between the import cost of high oil prices and reduced manufactured goods export opportunities, will find new evolving trade balances wherever they can. This will coincide with an increased need to seek out such trade on the part of Eastern Europe. This in turn could link trilaterally to oil producers, due to their own depressed circumstances in the current recession.

Any increase in non-CMEA trade that might result should not be attributed to a loosening of trade ties within CMEA. The Soviet attitude is not one of disengaging, as has sometimes been ascribed to this line of analysis. How can it be when three-quarters of the oil and all the gas imports in Eastern Europe are dependent upon the Soviet Union? But the incremental volume of oil is another matter.

Exports to Western Europe

Western Europe is perhaps best brought into focus at this point so that the full range of European options for the Soviet Union can be brought together. Total CMEA exports of energy to the whole of Western Europe and to the EEC in particular are shown in Table 4.1

Table 4.1 Council of Mutual Economic Assistance (CMEA) *estimated contribution of energy exports to total demand in Western Europe: 1985* (mto/mtoe)

	Oil*	Gas	Coal	Elec.†	Total
Western Europe total					
energy consumption	570	180	250	200	1,200
quantity from CMEA	80	27	15	1	123
% from CMEA	14	15	6	0.5	10

* Includes re-sales of non-CMEA source oil.
† Hydro and Nuclear only.

as they were before the current Five Year Plan. The proposed new gas trunkline deal will eventually double the export of gas. Statistically this might represent another 2 per cent of Western European energy requirements—although it may only occur when total energy consumption in Western Europe rises. Further increases in Soviet gas exports depend upon a revival in the world economy so higher gas output may be directed to internal CMEA use to free oil for export.

One element in the Eastern European pattern of oil exports has already ceased to be economic and principally affected Romania. It is the re-export of products from crude oil purchased in hard currency. CMEA in recent years has been the largest single external source of oil products for Western Europe. A sizeable part of this has been Romanian and some other Eastern European trade with OPEC crude oil suppliers and Western Europe product disposal: in effect the attraction of this has disappeared with the disappearance of refinery margins in the Western European market. Until a better balance re-emerges between OPEC crude prices, including their tax reference obligations to crude buyers, and the realization of product netbacks in the market place, the prospects for any refiner throughout Europe as a whole in this type of business look grim. One may add that Soviet crude, like any other oil on the spot market, will also have to compete with product netbacks at world market levels.

The significant point here is that the main impact of changes in the Soviet supply pattern affect Eastern Europe more than Western Europe. If by the time the Soviets plan to reach full export capacity of the Urengoi–Yamal gasfields they also plan to sell oil to *all* parties at world prices, with gas somewhat lower, this will put their customers in Eastern Europe under enormous economic pressure. It is the *rate* of change in world oil prices that is most damaging to importing economies. They tend to either over-react towards stringency or to rush for induced expansion, neither route having the structural time to work through without either inflation or stagflation resulting. If, in the longer term, the squeeze on oil prices induces real economies and real productivity compensation through expansion in trade, equilibrium will eventually be re-established. Can Eastern Europe manage with a lower percentage use of oil in its energy mix and still grow? I believe the answer is still no. The potential to economize on energy is large, but not the potential to substitute oil by other fuels. For example, except in Romania, the extraordinary high percentage of solid fuels in electricity generation—75 per cent to 90 per cent in the three northern countries—eliminates the option the OECD countries—and the Soviet Union—possess to remove oil from the electricity sector. Even if GNP growth is limited to 2 per cent annually, Eastern Europe needs to increase oil supplies at at least the same rate; that means another 10 to 15 million tons annually by 1986. This assumes some sustained recovery later in the current Five Year Plan period. There is a fair degree of flexibility already existing in the Soviet ability to supply both oil and gas. The current squeeze on oil imposed on Eastern Europe is clearly designed to press home both conservation and the substitution of Soviet gas—which as mentioned has been growing faster than oil for ten years now.

Should the Soviet Union achieve her hard currency export targets and her planned energy deliveries to Eastern Europe, the degree of flexibility mentioned may allow some variation in the mix of the oil and natural gas exports to each area. In other words, whilst the two markets are not interchangeable, the precise quantities of oil and natural gas delivered into each may be of secondary importance if the primary overall target level can be achieved.

It is unlikely, therefore, that a significant long-term impact is due to occur either in Western Europe or the Soviet Union as a result of changes in the exportable surplus of CMEA energy. The impact in Eastern Europe is much more severe and in the first half of 1982 the cutback they endured appeared to be equal in volume to the Soviet increase in hard currency oil exports. With decline in Romanian oil potential and the difficult road back for Polish hard coal during the current world recession, the danger of further economic implosion—or as in the case of Czechoslovakia, stagnation—is unpleasantly real. Czechoslovakia has just joined the list of countries ending the building of oil-fired power stations, so any further expansion of the solid fuel base in any of the Eastern European economies and the over-optimistic timing of the nuclear power programme may mean further slowdown, or greater dependence on the Soviet Union. Transfer of certain energy-intensive activities such as the production of alumina to the Soviet Union may help to improve the national energy coefficients in Eastern Europe.

Are major changes ahead?

For the Soviet Union itself, the 1986–90 plans detailed at the ECE and in the Ministerial statements are a portent of a major structural change in energy at the end of the decade. If these plans are achieved the Soviet Union would not only have become the leading oil producer and gas producer but also the premier coal producer in the world. Greater emphasis is placed on coal expansion in the second half of this decade than on oil and gas combined. Oil is under pressure—it may even have peaked during this current Five Year Plan. Problems with further gas penetration may be a principal factor in the relatively slow increase in gas projected after 1985 as compared with the current plan period. The nuclear power programme, highly concentrated in the European CMEA, the Soviet Union included, will be hard pressed by the time factor, and at the base of all this energy development will be Asiatic coal and the crucial question of the rate of progress of 'coal by wire' in a nation spanning eleven time zones. Will it be possible—and if not, will it actually be necessary?

The answer to this question depends almost wholly on the answer to another—will the CMEA countries still demand roughly twice the energy input of Western European countries for the same economic output? In March 1982 the BBC reported a Moscow radio broadcast as stating that the potential for fuel saving in CMEA was 900 million tons of standard fuel. That is the same problem, quantified. There can be little doubt of the sincerity of the Soviet policy to remain self-sufficient in each form of energy. But measures to improve the economy on the demand side of the equation must find a complement in further production increases. The position is more urgent in Eastern Europe where self sufficiency is out of reach and economic viability itself may be the central issue. But by 1990 the Soviet Union itself will be facing the same issue. There is little doubt that the inclusion of the subject of energy efficiency in his first major speech as First Secretary by Yuri Andropov was highly significant, and that after a brief period under Chernenko, that theme will re-emerge as Gorbachov reasserts the new efficiency drive.

Meanwhile the trade balance should not present any undue problems for the Soviet Union, unless world oil prices become unduly unstable. But they *do* represent severe problems for Eastern Europe with its singular need for more oil. The Soviets have demonstrated in recent years their ability to switch oil exports from East to West Europe and to put pressure on Eastern Europe to take more gas. They have also made great strides in substituting gas for oil in leading energy intensive industries in the Soviet Union and in the use of combined heat and power. Moreover, they have shown that where oil is in a non-substitutable sector such as gasoline, they will not hesitate to use *prices* to curb excess demand. But in future further oil saving gains must largely come from inside the Soviet Union itself.

The Soviet trade pattern is one in which exporting 10 per cent of their oil and gas output pays for imports of food, machinery and raw materials in about equal measure—essentially to assist Soviet agriculture, heavy industry and energy development. There is a major prize to be gained within the Soviet Union by improving the agricultural outlook either in productivity or distribution (i.e. improving the delivered percentage of the crop), although it is perhaps not sufficiently appreciated by the world at large that grain output rose 50 per cent during the Brezhnev period. A 10–15 per cent improvement in agricultural efficiency would be of as much advantage to the balance of payments as the total current volume of gas exports. Coal also stands to gain considerably from the present drive for efficiency. Therefore any economic feedback from international trade may provide a non-market system with the opportunity to evaluate *demand management*. The main impact is not whether the Soviets achieve all their energy *production* targets by 1990. Rather it is a question of where they can go from there if *efficiency in demand* does not take a decisive step forward.

The significance of this question for Comecon oil prospects is crucial. If energy efficiency can compensate for, i.e. offset delay in the supply of, any of the major primary energy programmes in the 1980s and if natural gas continues to supplant oil, then the pressure on Soviet oil production and Soviet ability to export will ease more significantly than by any other means. But there is a limit to the extent that these measures can be provided by Eastern Europe's oil-deficit. It is the Soviet Union, in the long term, that will have to provide the means. At present internal Soviet oil demand for unique, i.e. non-substitute, purposes, is about 3 mb/d out of total oil use of 9 mb/d and total energy use of 24 mb/d. The other 6 mb/d of oil demand is potentially substitutable. Oil exports to non-CMEA markets represents only about one-quarter of this volume of oil and 5–8 per cent of total energy consumption in the Soviet Union. Put in this way it does not sound too formidable a target to find ways of maintaining an adequate volume for hard currency export, increasingly assisted by sales of natural gas.

Before the target can be reached, however, the main problem is not one of adequate energy resources so much as the ability to achieve means of enabling the internal Comecon market to conserve oil. Essentially this refers to the Soviet Union itself rather than Eastern Europe because of the former's much greater size and flexibility. Technically it appears to be possible to make efficient substitution of other energies for oil. This would have to be part of an acceptable evolving social structure within the tenets of the existing politico-economic orthodoxy. The physical potential to do so exists. Also the drive to efficiency throughout the economy continues to be debated at all levels and in many activities including energy, still all this leaves open the question whether it will be possible to deal successfully with the demand aspects of the energy balance before the supply potential becomes seriously extended.

Whatever the outcome may be one thing emerges very clearly: the vast majority, perhaps 90 per cent, of Soviet energy exports are linked to Europe, either Eastern Europe or Western Europe. Solutions for part of Eastern Europe's future oil development may lie with non-Soviet suppliers if they can solve their particular problems of trade. But the main bulk of their supplies will continue to come from the Soviet Union. It is only in incremental terms, and only in oil, that this growing tendency for Eastern Europe to look elsewhere is likely to be significant. 'Elsewhere' must mean *non-European* sources of oil. We are not envisaging a Western to Eastern European oil flow of any significance.

Leaving Eastern Europe aside, Soviet energy supplies to Western Europe comprise about 10 per cent of Western European demand. This outlet to the Western European market is almost certainly going to represent the vast bulk of total CMEA energy exports for the foreseeable future. In case this assertion sounds exaggerated, let us briefly examine the alternatives to Western Europe as outlets for CMEA energy exports. The other hard currency possibilities are North America and Japan. For a start one may rule out natural gas projects—such as the once vaunted 'North Star' for Siberian gas by LNG carrier to North America, or Yakuts gas via pipeline (2,000 miles) to the Pacific and then by LNG carrier to California. The costs of such projects and the improved prospects for gas discoveries within North America, together with very long lead times, rule out any such prospect within this century. Oil trading with the United States has been, and is likely to remain, negligible. Exchange deals whereby Venezuela supplies Cuba, a full CMEA member, in return for Soviet supplies to a third party—probably in Europe—are little more than convenient geographical exchanges. Polish coal, if it fully recovers momentum, may continue to reach Western hemisphere outlets in small quantities. But it is at a locational and cost disadvantage to regional sources of supply. The Japanese position has already been discussed, important as Japanese funding is to the development of the Soviet Far East, the relatively small Far East deposits of oil and gas in the Sakhalin region and the eastern exports of coal remain at less than 1 per cent of Soviet production of each fuel and an almost equally minuscule proportion of total Japanese requirements.

In the developing world, small quantities of Soviet oil and Polish coal continue to be supplied but direct sales of oil have declined over the past ten years and are now almost solely the result of trade acquisition of foreign oil–e.g., from some OPEC states–by the Soviet Union being on-traded in this way instead of being physically imported into CMEA. This arises out of technical assistance to some national energy corporations who repay their advisers in kind–or under multiple reciprocal economic bilateral deals between the Soviet Union and Third World countries where some oil is obtained by the Soviet Union in the trade balance for goods and services exported. In some cases the oil may enter Eastern Europe but rarely, if ever, does it get imported into the Soviet Union, although in oil-deficit Eastern Europe it is clearly more significant.

There remain the Asiatic nations bordering upon the Soviet Union between Turkey and Japan. Both Turkey and Greece are OECD members. In this sense they are included in Western Europe. Iran and the IGAT pipeline supply of gas produced in the Iranian oilfields represents one of the few imports of energy into the Soviet Union. It may be coupled with imports of natural gas from Afghanistan to Dushanbe, since neither has any alternative outlet. Before the oil price rise in 1979 following the fall of the Shah, 'IGAT II', to carry associated and non-associated Iranian gas to the Caucasus in return for Soviet gas supply to Czechoslovakia and westwards, looked to be a major breakthrough in providing more Middle East gas (back to back with Soviet gas) to Central Europe. Now, the massive Soviet gas developments and the 500 per cent rise in asking price of Iranian gas appear to have closed the prospects of such a deal for the foreseeable future. Some more Afghan gas is being developed for the Soviet Union but quantities are small–less than 2 per cent of Soviet gas production–and only exist because of the locational attraction to both parties in the region.

Mongolia (full CMEA member) and North Korea both obtain their modest oil needs from the Soviet Union on a scale compatible with railroad delivery. This leaves China. The complementarity of Soviet Union–China energy patterns is often reviewed and as often found to be negative. China's coal production rivals the Soviet Union. Chinese oil production, 2 mb/d is low compared with her potential demand, but all attention is now focused on her promising offshore oil areas, as yet untapped, but where, with Western help, commercial oil has been announced in 1983. There is no similar Chinese natural gas development as yet, but gas from coal-producing provinces must be a priority for the future, long before turning to the Soviet Union whose gas deposits are mainly linked westwards and lie thousands of miles from the populous areas of China. In the same way, even a technical breakthrough in long-distance high-voltage electricity from the coalfields of central and eastern Siberia, that might, one day, reach towards India and China, will have to compete with the substantial local coal-based production of power in each of these major markets. The Soviet trade prospect reviews all these and returns, as at present, to its European sphere of production and outlets, both domestic, CMEA and hard-currency earning-exports.

Western Europe, however, is in a very different basic energy position to Eastern Europe. The bulk of its supplies comes from non-Soviet sources. Also it enjoys a much higher degree of flexibility between primary energy sources and between countries. While the implications of this speak for themselves it is fair to add that, in total, Western Europe remains more energy-import dependent than North America. It is therefore clearly not possible for Western Europe to regard susceptibility to import dependence as a factor that can be curbed or dealt with to the same degree open to the United States or Canada. It is more analogous to the options open to Japan. Dr Paul Frankel, in a memorable phrase, described this as 'the diversification of insecurity'. In this sense Western European dependence on Comecon supplies of energy is never likely to approach the level of that on world oil supplies or to become as concentrated as are Soviet energy exports upon one region, i.e. the Western European market. East–West European energy trade, therefore, must be seen in a global context and not simply in the purview of a European perspective.

Chapter 5

North sea oil and gas and energy security for Western Europe

Øystein Noreng

North Sea oil and gas in a West European perspective

Oil and gas are 'political' commodities in the sense that secure supplies of energy have a fundamental economic and strategic significance for the consumer countries. Even fairly minor disruptions of supplies can cause disorder in the market and major price rices. The alternatives are keeping large and expensive stocks or reducing energy demand through a reduced level of economic activity, also at high economic and social cost. In political terms dependence upon imported oil and gas may restrict freedom of action in relation to major suppliers. Against this background, secure energy supplies, preferably from domestic sources, have an additional value, a 'political rent'. Though not always expressed in money terms in the marketplace, it may nevertheless have other forms of compensation. In this perspective, the North Sea oil and gas resources have a particular significance for Western Europe. They reduce dependence upon external energy supplies and consequently, exposure to economic and political risks. However, this blessing is not without certain problems to Western Europe.

Henceforth, the term 'North Sea oil and gas' will be used to cover the United Kingdom and Norway, the two principal offshore producers. This definition therefore excludes Danish, Dutch and West German offshore oil and gas as they are of minor significance.

Assets and problems of North Sea oil

For Western Europe, North Sea oil and gas have been important in reducing dependence upon external energy supplies since the early 1970s. Still, Western Europe remains the world's largest importer of energy and is likely to remain so for the foreseeable future. Great progress has been made in reducing dependence upon foreign supplies from 64 per cent in 1973 to 44 per cent in 1982. Domestic energy output has increased from 450 million tons of oil equivalent (mtoe) to 682 mtoe over the same period, a net increase of 232 mtoe. North Sea oil and gas represents close to two-thirds of this increase, 153 mtoe, i.e. 125 million tons of North Sea oil and 28 mtoe of North Sea gas: 110 mtoe came from the United Kingdom, 43 from Norway. The remainder of the increase comes essentially from nuclear energy. Thus, since the tables were turned in the world energy markets in 1973, North Sea oil and gas have been by far the most important element in increasing Western Europe's energy independence, and consequently its energy security.

In 1982, Western Europe's total energy requirements were 1,217 mtoe, slightly less than the 1,241 mtoe of 1973. Oil demand was 601 million tons, gas demand was 174 mtoe. In this perspective there is an evident complementarity between what the North Sea has to offer and what Western Europe needs in terms of oil and gas. The advantages of North Sea oil are obvious. The fields are located close to the major markets and trading centres of North West Europe. They are situated in countries with a high degree of political stability and close economic and political ties with the rest of Western Europe. Both the United Kingdom and Norway are members of the OECD and NATO; the United Kingdom is a member of the European Community, with which Norway has a trade agreement; the United Kingdom is a full member of the International Energy Agency (IEA), while Norway has an associate membership. No other major oil and gas producers, except the Netherlands, have as close political ties with the West European importers as these two countries. Furthermore, transportation of North Sea oil and gas to the major West European markets is easy (mostly by pipeline). Thus, the risk of interruptions is minimal. Against this background, it is evidently preferably to depend on North Sea oil and gas rather than on supply from the Middle East, North Africa or the Soviet Union. However, as mentioned earlier, this advantage has a premium.

There are a number of problems related to North Sea oil and gas, apart from costs, that will be dealt with later. As Ian Smart points out (cf. Chapter 8), it is somewhat simplistic to view Western Europe as a single unit in terms of energy. The region consists of twenty-two independent countries, with a corresponding number of different economies and requirements as well as governments and preferences. The division between exporters and importers of energy is especially important. They trade in commodities in a market where the driving forces and balances are essentially set internationally, i.e. outside their control. Thus, in the intra-West European oil and gas trade, there are important differences between exporter and importer interests that seldom are overcome through political agreement. As will be elaborated later, secure supplies have a cost, partly because of the political rent mentioned above. Also, as Ian Smart points out, since North Sea oil and gas are generally expensive, there is an economic cost if the consumers rely on expensive and secure domestic supplies rather than less expensive and less secure external supplies. Costs and risks have to be weighed against one another. The outcome is partly a result of political perceptions, partly a result of economic requirements.

The North Sea potential

Since the discovery of gas in Groningen in the Netherlands around 1960, estimates of the petroleum potential in the North Sea have varied tremendously. The northern part was considered less promising until oil was found. More recently it has been considered rather promising. A resource becomes an exploitable reserve of economic interest as a function of prices and technology. Since the early 1960s the price of oil has jumped twice, and technology has greatly progressed. Inevitably, North Sea oil and gas have become more attractive. The total resource base is still fairly uncertain, because exploratory drilling is much more difficult and expensive at sea than on land. Thus, reserve figures are essentially conservative. There has been a gradual escalation of estimates. The British potential is better known, as exploratory drilling has been more intense on the British continental shelf than on the Norwegian side. Total recoverable oil reserves in the British area are estimated to be in the range of 2,100–4,300 million tons, and the maximum recoverable gas reserves to be about 1,400 billion cubic metres.[1] Norwegian reserves are more uncertain, but total expected recoverable reserves south of 62° N are estimated at 4.7 billion tons of oil equivalent.[2] North of 62° N any estimate is pure guesswork, given the limited amount of drilling so far. But the area covers about 85 per cent of Norway's continental shelf, and resources here could be significant by international standards.

These figures include potential reserves at water depths of more than 300 metres. In the United Kingdom remaining reserves at water depths of less than 300 metres in present discoveries are about 800 million tons of oil, with a similar figure for deeper waters. Remaining reserves of gas in present discoveries are in the range of 900–1,200 cubic metres. In Norway, the remaining reserves of oil and gas in present discoveries are estimated at 2,500 mtoe. Even if total North Sea oil and gas reserves may be high, they are to be found and developed under increasingly difficult and costly conditions. Without venturing into deep waters, the United Kingdom has a reserve/production ratio for oil of about 8:1, for gas about 30:1. The corresponding Norwegian ratio for oil and gas is about 35:1.

The higher level of exploration in British waters means that most of the larger fields have already been found. Probably only relatively small oil and gas fields will still be discovered. A qualification should be made for the deeper waters west of Shetlands and the Orkneys. Here, little drilling has taken place so far. By contrast, the historically lower level of activity in Norwegian waters means that chances of finding large fields are better even if some of them are in deep waters. An obvious example is the Troll field. It is considered to be so far the world's largest offshore gas field (at 350 metres of water). It has a potential of close to 2,000 billion cubic metres which are not included in the reserve estimates mentioned above.

Oil and gas policies and supplies

There are both contrasts and parallels between British and Norwegian policies toward the domestic oil and gas resources. The major difference is to be found in depletion policy. The United Kingdom has opted for a quick extraction policy, especially for oil, whereas Norway has opted for a much slower depletion, stretching out the life-time of reserves. The origins of these policies stem from deliberate political choices. They also have their background in economic circumstances and in quite different economic performances.

The British oil and gas were found in a country that since 1945 had had a sluggish and uneven economic growth, with a chronic balance of payments problem, and a generally low rate of capital investment. During the same period Norway had enjoyed a fairly high and remarkably stable rate of economic growth, with no serious balance of payments problems and an extremely high rate of capital investment.[3] In the United Kingdom the rate of investment increased slowly from about 16 per cent in 1950 to about 22 per cent in the mid-1960s.[4] The Norwegian rate of investment has generally been between 30 and 35 per cent of Gross National Product since 1945. These figures indicate differences in availability of capital in the two countries, and in the need for money.

Deciding the depletion rate of a finite resource is an investor's choice.[5] Oil left in the ground can only be of economic value to the owner to the extent that it increases in value, and that its value increases at a faster pace than the perceived cost of money. Thus, in hindsight, the fast pace of depletion of the British reserves may be explained by a high discount rate, with the British government judging the utility of petroleum revenues in the near future as high, and higher than that of leaving oil in the ground, even if it should appreciate in value over the time horizon considered. The Norwegian choice was rationally quite the opposite. The Norwegian government had few reasons in the late 1960s and early 1970s to use a high discount rate. Consequently, oil left in the ground could be seen as a more attractive investment. Furthermore, Norway with a population of only 4 million, as opposed to the 56 million of the United Kingdom, had reasons to fear overheating the economy by large-scale petroleum revenues.

Over time, there have been some changes in the depletion policies of both countries. The United Kingdom at first embarked upon an aggressive depletion policy. After some hesitation under the Labour government, the Conservative government, after 1979, firmly decided for a quick depletion. In practice, this meant net oil exports for most of the 1980s. Norway in the early 1970s opted for a fairly high level of activity, which later could have given much higher volumes of output. After the oil price rise of 1973–4 there were strong fears of economic overheating and uncontrolled social and economic change. The exploration activity was curtailed, and a policy of going slow was initiated. Already by 1978 it was evident that capacity to spend petroleum revenues was much higher than anticipated in 1974. Furthermore, the international economic recession which hit Norway's traditional industries, meant that Norway's problem was no longer that of potential economic overheating. Rather it was one of maintaining a satisfactory level of economic activity. Exploration was stepped up. The results were positive. They showed a much larger resource base than previously anticipated. This in turn permits a higher level of investment, activity and output, while still maintaining a high ratio of reserves to extraction. However, Norway's change in petroleum policy in the late 1970s can largely be explained by changing international economic circumstances.

There are significant differences in the organization of the petroleum industry in Norway and the United Kingdom. They touch primarily on the role of state participation. Even in 1969, Norway introduced a provision for state participation in all new licences. This led to the establishment of a national oil company, Statoil, in 1972. In 1984, state participation was split between the government and Statoil, the two together representing between 50 and 85 per cent of all new licences. By contrast, in the United Kingdom, state participation was only introduced in 1975, with the establishment of BNOC, the British National Oil Corporation (broken up in 1982). In 1985, the BNOC was exclusively a government oil trading organization. It enjoys a remarkable lack of flexibility which makes it highly dependent upon the spot market. In the United Kingdom, the British Gas Corporation had a purchasing monopoly for gas until 1982. This caused depressed prices for producers and low ones for consumers, stimulating the introduction of natural gas in the British energy market. In Norway, all natural gas is exported.

In the United Kingdom gas extraction commenced as early as 1967, reaching an annual output of 33 billion cubic metres (bcm), by 1983. Oil extraction was started in 1976 and reached 115 million tons by 1983. By this time, domestic consumption was 42 billion cubic metres of gas and 73 million tons of oil. This made the United Kingdom a net exporter of hydrocarbons, with 42 million tons of oil exported and imports of 9–10 billion cubic metres of gas.

In Norway, oil extraction commenced in 1971, reaching 30 million tons by 1983. Gas extraction began in 1977, and reached 23 billion cubic metres in 1983. In that year, domestic consumption was 8 million tons of oil, making Norway a more important net exporter of hydrocarbons than the United Kingdom.

In both countries, future oil and gas output is uncertain. In the United Kingdom, the recent petroleum tax reductions have stimulated exploration and development. With essentially smaller fields yet unexplored, the cost curve is bound to rise. The projected decline in output will probably only be slowed down but hardly reversed. On balance, however, the United Kingdom may remain a net oil exporter throughout this century. Uncertainty about future gas output is at the roots of the controversy over the deal to import Norwegian gas from the Sleipner field. Possibly, gas output could rise significantly in the 1990s and then decline again after 1995, opening up additional markets for imports. Future gas imports need not come only from Norway, they might also come from the Netherlands or even the Soviet Union. The controversy over gas imports is also linked to a struggle between the British Gas Corporation and the domestic gas producers (i.e. to a large extent the United Kingdom based oil companies) over the division of economic rent.

The outlook for Norwegian output is more uncertain. The Norwegian government faces some important policy choices concerning both its level and the balance between oil and gas. A recent report by a Royal Commission recommends an accelerated level of development, essentially in order to stimulate Norwegian industry.[6] Translated into volumes this could mean an output of 100–200 mtoe by the year 2000, or practically twice the present figure. The trend toward a much higher level of output is also sustained by Norway's need for revenues in the 1990s, in view of the poor performance of export industries.

In recent years, there were more discoveries of gas fields than of oil fields. This could mean that Norway will have to develop some of the large gas fields in order to sustain the investment and reach a higher level of output. However, as will be discussed later, the West European gas market may remain fairly depressed for several years, with gas prices lagging considerably behind oil prices. Thus, going for oil seems more interesting for Norway. Therefore, if more oil fields should be found in the near future, Norway is likely to develop them first.

To sum up, the total North Sea oil output is at best likely to remain even throughout this century, with a possible slight decline.

Gas output, however, is likely to increase and will affect the West European market.

North Sea oil and gas in the West European market

The advent of North Sea oil and gas in the 1970s broke a historical trend of declining self-sufficiency in energy for Western Europe. Since 1973, this trend has been reversed in a double sense. The growth of energy consumption has been largely checked. Total primary energy consumption in 1982 was less than in 1973. This is due to increased energy efficiency and a sluggish economy. Increased energy efficiency has necessitated large-scale investment; the decline in demand has also had a high economic and social cost. After each of the oil price explosions, the West European economy and energy demand receded. During the period of recovery, essentially 1976 to 1979, energy consumption in the region rose at an annual rate of 3.1 per cent. This might well repeat itself. After that, energy consumption declined in 1980–3, but could start increasing again if the economy picks up. It should be noted that the oil price rise of 1979–80 was followed by a period of high real interest rates, contrary to the period after 1973–4. This may have discouraged some investment in energy conservation.

As discussed in other chapters we must, therefore, assume that in the future, Western Europe's energy demand may be even more uncertain than North Sea oil and gas output. A projection based on current trends could indicate that energy consumption until 1990 will increase at an annual rate of slightly more than 1 per cent and then practically stagnate in the 1990s. Total energy consumption in 1990 will be somewhat above the 1979 level, around 1,350 mtoe, reaching 1,450–1,500 mtoe by the year 2000. The position of oil will recede, relatively and absolutely. The share of natural gas will remain fairly constant, implying a modest increase in volume. The role of nuclear power and coal will increase substantially. In this perspective, the North Sea will be of value in gradually raising the level of self-sufficiency in oil and in at least maintaining the current level of gas output in Western Europe.

Should, however, economic growth pick up again in the mid-1980s, total energy consumption would increase quite vigorously with a commensurate growth of oil demand and oil imports. As Hans Maull points out in his contribution, this could quite easily contribute to triggering off a third oil price explosion, possibly in conjunction with political events in the Middle East at a time when the oil market is particularly vulnerable, e.g. in a situation of quickly rising demand but with low inventory levels. This would again plunge the West European economy into a recession. At the same time, it might stimulate the restructuring of the economies and of the energy sector. Total energy demand by the year 2000 might then be in the vicinity of 1,800 mtoe, following the OECD low demand scenario.[7] However, the shares of the residential and transportation sectors would be higher than currently projected, at the expense of the share of the industrial sector. Oil would be much more restricted to its premium use as a fuel in transportation and as a petrochemical feedstock than is the case today, but the use of natural gas would be much greater than currently anticipated. In this scenario, North Sea oil would be even more valuable to Western Europe. Its light quality provides an easier refining basis for premium products than the heavier crudes imported from the Middle East. Natural gas from the North Sea would be especially important.

This alternative future may not be unrealistic. But the problem is that it has little impact upon today's energy planners and decision makers. There is a contrast between the long lead times of energy projects and the short time horizons of energy planners and decision makers. They often tend to take the immediate past as the basis for the future. Just as the oil price explosions of the 1970s were followed by a run for energy deals, the current depressed energy market leads to a strong reluctance to engage in new energy deals. Important consumer countries that a few years ago took drastic measures to obtain energy security are today facing an embarrassing and costly surplus of energy that is projected far into the future.[8] It also discourages investment in new energy projects.

As an economically marginal area of development the North Sea is strongly affected by price fluctuations in the world energy market. The viability of investment projects required to bring additional supplies of secure energy to Western Europe is dependent upon the demand and the price in the marketplace, unless special measures are taken to protect new projects. This is the economic background of the political gamble over West European gas.

North Sea gas and West European security

The projected decline in continental gas output means that substitute gas supplies must be found to sustain current consumption, utilizing the infrastructure in place. If gas demand should increase (for example in a different economic and political context in the 1990s), additional supplies would have to be found outside the major consumption area of continental Western Europe and the United Kingdom. Current contracts cover projected demand until 1990. Thereafter, there is a projected discrepancy between probable demand and contracted supplies. For the late 1990s this discrepancy is about 60–70 mtoe, or close to a third of projected demand. If gas demand should increase more than currently envisaged, the discrepancy could be substantially higher. The choice for energy planners in governments and companies is thus between contracting larger gas supplies, substituting other fuels, essentially nuclear power and coal, or raising prices to curb demand, or a combination of the three.

Apparently contracting larger gas supplies is the easier solution. Raising consumption of coal and nuclear power could be politically difficult: raising energy prices further could have an undesirable economic effect. Western Europe is surrounded by large untapped gas reserves, in the North Sea, in North Africa and in the Soviet Union.

The North Sea gas reserves are indigenous to Western Europe. They represent an especially attractive source of supply, at least from a security point of view. However, North Sea gas also faces stiff competition in the West European market. For our time horizon, i.e. the rest of this century, large-scale overseas gas supplies, whether from West Africa, the Middle East or Canada, can be ruled out. The reasons are high transportation costs, whether by LNG tanker or by pipeline, and long lead times. This limits competition to North Sea gas essentially to gas from Algeria and the Soviet Union.

Algeria has large reserves of natural gas for which Western Europe is the obvious market. Capacity in the current pipeline system to Italy is limited to 12 billion cubic metres (bcm) a year. It could be raised to 18 bcm a year by enhanced compressor capacity. Any further growth of gas exports would imply either transportation by LNG tankers (which is costly) or the construction of another pipeline, possibly to Spain (which requires a long lead time).

The Soviet Union, as Anthony Scanlan points out (cf. Chapter 4) obviously has a reserve base sufficient to sustain substantial new gas exports to Western Europe. Soviet flexibility in gas exports is enhanced by the fact that exports only make up a fairly small proportion of total volume extracted. Excess capacity in the current and new pipelines could enable the Soviet Union to raise exports by 20 bcm a year at low marginal cost. The Soviet Union could be somewhat restricted in her gas exports to Western Europe by the need to substitute gas for oil domestically and to raise gas exports to Eastern Europe.

Both in Algeria and in the Soviet Union costs of extracting natural gas are substantially below that of the North Sea. Transportation costs, due to larger distances, may be somewhat higher, but they do not make up for the former. Pricing may have made Algerian gas in recent years less of an obvious first option to West European consumers. Soviet pricing policy has been more moderate, at least for incremental volumes within the present pipeline system. This made Soviet gas more attractive. The Soviet Union also has a record of supply reliability. The need for foreign exchange in many ways makes her dependent upon gas exports. It could induce her to step up natural gas exports to Western Europe. Thus, the competition seems to run essentially between Norwegian and Soviet gas.

Politically, Norwegian gas may be the more attractive to West European consumers, but economically Soviet gas may be the more attractive. Dependence upon Norwegian gas presents no apparent problems to the West European consumers, but the economic costs are high. Gas distribution companies tend to give economics a higher priority than politics. Governments are likely to set the balance according to constantly changing circumstances. East–West relations are an important ingredient in these considerations. In commercial terms, a high degree of dependence upon Norwegian gas may not be judged entirely desirable from a gas distribution company point of view. In the offshore Norwegian petroleum industry there is a risk of accidents. There is also a potential problem of labour unrest which does not exist in the Soviet gas industry so far. Furthermore, it may give the Norwegian sellers a bargaining power in the marketplace that could enable them to push up the price of gas. The Soviet alternative may be less attractive politically, but economically and commercially more advantageous to companies. Governments are better placed than companies to perceive political risks.

The competition between Norwegian and Soviet gas in Western Europe is basically one between two oligopolists. Structurally, the West European gas market is far from the competitive market described in economic theory. It may be described as an oligopoly, with few sellers on the supply side, and a single buyer on the demand side (various agreements between the various continental gas companies). If the United Kingdom is included, there is a 'duopsony', i.e. two buyers on the demand side. Oligopolistic markets are characterized by potential instability and by the ability of each of the participants to influence the outcome, and thus their own situation. In such a market, change may be marked by discontinuities. Adjustment is carried out by conscious actors and not by impersonal market forces. Thus, a long-term strategy is essential in the West European gas market. This is valid for the buyers as well as for the sellers. As both volume and composition of future gas demand are uncertain, the buyers find it risky to contract large incremental volumes of gas at high prices.[9] At the same time, delays in contracts and lower prices may defer the development of gas fields in the North Sea. This leads to a risk that if demand rises more than anticipated the gap is filled by supplies from other, external sources, prices are likely to rise. The shorter lead times and lower costs put the Soviet Union in a favourable position as a future incremental supplier. This again could increase the Soviet share of the West European gas market to the detriment of the Norwegian share. Such a perspective is less troublesome to the companies than it is to the consumer governments.

For Norway, the key issue today is whether to expand gas output over the next ten–fifteen years and at what pace, or whether to opt for an accelerated development of oil. The outcome largely depends upon economics. The Troll field and possible offshore North Norwegian fields, from which an accelerated output of gas would largely have to come, are in the current situation at best marginal from an economic point of view. At current prices their development might require a modification of the taxation system. This, however, is seen as undermining the bargaining position of Norway in relation to the oil industry. But if Norway chooses the oil option, it could become a more important supplier to Western Europe (or rather to the OECD area) as the major market for Norwegian oil currently is the East Coast of the United States. Following the oil option, Norway would not push to expand its share of the West European gas market. Rather, it would wait for better times, when the consumers would demand the development of more Norwegian gas, at prices more acceptable in a Norwegian perspective. This would leave a great part of the incremental gas demand in Western Europe to the Soviet Union with the qualification, however, that the consumer desire for diversification might lead to a certain demand for Norwegian gas resulting from new contracts for Soviet gas. Thus, each additional volume of Soviet gas sold to Western Europe might prepare the market for a smaller volume of Norwegian gas, possibly at higher prices.

Norway and the Soviet Union: convergence and competition

For the main competitor, the Soviet Union, the key issue is apparently how far the West European gas market can expand and at what prices, and what share will be tolerated for Soviet gas. Facing a slowly declining output of oil, the Soviet Union has an interest in building up gas exports. Throwing large additional volumes into the market could easily depress the price and reduce the netback value. Furthermore, to the extent that there is a ceiling beyond which the Soviet market share is not permitted to expand, the Soviet Union has little interest in reducing prices in order to stimulate demand and expand sales. In this perspective there could be a Soviet interest in letting Norwegian gas exports to Western Europe expand, in the expectation that each additional contract for Norwegian gas could prepare the market for additional volumes of Soviet gas.

Against this background Norway and the Soviet Union, as the two leading future exporters of natural gas to Western Europe have both competing and complementary interests. On one hand, there is competition for contracts in the West European market. On the other hand, there is a complementarity of interests: neither country would find it useful to launch large new volumes of gas into the market at low prices, but either could have an interest in the other party doing so. They would benefit from the consumer desire for diversification without having to bear the economic cost or facing the political obstacles of being the driving force in expanding the gas market.

In this perspective there is ground for understanding, if not outright agreement, between Norway and the Soviet Union. They might be interested in elaborating a common strategy in relation to the West European gas market. However, obvious political obstacles prevent such a commercial alliance between a smaller NATO country and the Warsaw Pact superpower. Thus, East–West considerations have a profound impact upon the organization and behaviour of the supply side in the West European gas market.

The West European gas supplies have recently become an object of both East–West and West–West politics. In the debate as to whether and to what extent Western Europe should depend on Soviet gas exports, Norway has become an important point of reference rather than an active actor. American hopes and Soviet fears are that Norway will be able to 'take' the Western European gas market from the Soviet Union. By providing alternative supplies she would effectively bar the entry of more Soviet gas to Western Europe. The Reagan Administration has been leaning on Norway to provide alternative gas supplies. However, it has hardly been aware of the constraints that have to be overcome, both of a technical and a political nature. As pointed out earlier, Norwegian gas has higher costs and longer lead times than Soviet gas. Furthermore, to provide Norwegian gas in quantities and at prices that can effectively compete with Soviet gas would require a reversal of Norwegian depletion and taxation policy.

This politicization of the Western European gas market presents an awkward dilemma to Norway. Norway might benefit from American pressure to bar the entry of more Soviet gas because it could then sell her own, more expensive gas. This is essentially the reason why at the IEA, in spring 1983, the United States and Norway in their wish to impose a ceiling on the Soviet market share jointly opposed the other members. This expressed a convergence of American foreign policy interests and Norwegian energy policy and commercial interests. Norway expressed satisfaction that the report on energy security actually mentioned the Troll field, hoping this would facilitate its development. However, the IEA report so far led to no commitments from companies to buy Troll gas. This shows that from a company point of view the security of Norwegian gas as opposed to Soviet gas may be of little practical importance.[10]

For Norway it is of limited practical value to line up with the Reagan Administration in order to market her gas, and to let the prospects for exporting gas be tied to the ups and downs in East–West relations. It is contrary to the traditional petroleum policy and foreign policy of Norway, and it might backfire. It might compromise relations with the Soviet neighbour, and it might compromise the prospects for marketing Norwegian gas if there should be a thaw in East–West relations. Finally, it might compromise Norwegian petroleum policy and invite additional American pressure on depletion and taxation issues.

The economics and politics of energy security

The notion of 'energy security' implies an imperfect market. In a fully competitive market, with no barriers to entry, the cutback of one supplier will spontaneously be offset by the entry or increased volumes of another supplier. Under such ideal circumstances security of supply is an irrelevant problem. Barriers to entry on the supply side and the propensity of suppliers to reach at least tacit understanding usually made most energy markets less than perfectly competitive. Oligopolistic structures dominated. Oligopolistic markets are by their nature unstable, as the actions of one supplier can affect the outcome. Economic, political and technical factors prevent the cutback of one supplier being automatically filled by another. They make security of supply a real issue.

As the international energy markets have a propensity for cyclical movements, with short periods of scarcity alternating with long periods of glut, this is only of theoretical interest most of the time. Sometimes it can, however, suddenly become acutely relevant. For example, during the decade of the 1970s, the oil market was in turmoil during only 18 months out of 120 (October 1973–February 1974, plus December 1978–December 1979). Thus, for 85 per cent of the time the problem of supply security seemed irrelevant. However, during the periods of acute relevance it cost the consumers two major price rises and some uncomfortable political problems.

During the periods when the markets were calm it was economically advantageous to rely upon spot or random purchases at low prices. This was, for instance, Sweden's policy in the late 1970s. However, for the 15 per cent or one-sixth of the time, it was certainly economically highly advantageous to have secure oil supplies. But the balance is not only a question of price differences and discount rates. Insecure energy supplies in times of crisis also seem to push buyers into some degree of irrational behaviour. During 1979, the year of perceived shortage and rocketing oil prices, more oil was extracted and traded and consumed than in any other year in history. The perception of scarcity induced consumers to hoard. This increased overall demand and pushed up prices. In hindsight it appears that a more rational behaviour in 1979 could have led to a less dramatic price rise in oil. In this perspective energy security has also to do with behaviour and politics, strategy and management. Thus, the IEA report is correct in pointing out that a Western European nightmare would be the simultaneous cutback of Middle Eastern and North African oil as well as of North African and Soviet gas. Irrespective of whether such a simultaneous cutback is likely, the United States is probably correct in assuming that Western Europeans' reactions in the face of such a crisis may well be unpredictable.

Against this background North Sea oil and gas have a political significance in Western Europe. A policy which substitutes expensive Norwegian gas for cheaper Soviet gas may in the first instance increase the real price of energy to consumers. Subsequently, however, it may help insure the Western Europeans against future supply cutbacks and improve their bargaining position in relation to external energy supplies.

The premium eventually given to Norwegian gas over Soviet gas involves both governments and companies. It is discretionary as the tolerance level of the Soviet market share in the Western European gas market is closely linked to the state of East–West relations. This again is linked to trade as well as to Soviet behaviour in Eastern Europe and other parts of the world. So far, the conventional wisdom has been that Norway, eventually together with Algeria, will play a secondary role in the West European gas market, with outlets for Norwegian gas largely depending upon the desires of the continental buyers to diversify after bulk purchases of Soviet gas. In this perspective, the expansion of Soviet gas exports was seen as the driving force in the West European gas market. It led to smaller shares of the market for more expensive Norwegian and Algerian gas.

Recent discoveries of more gas in Norway and political considerations could change this picture. With a reserve/extraction ratio for gas possibly more comfortable than that of the Soviet Union, Norway has an increasing interest in giving a priority to volumes and market shares rather than to high gas prices. To the extent that Norwegian gas takes a larger share of the market, the bargaining position of the continental buyers in relation to the Soviet sellers will improve. Supplementary volumes of Soviet gas might therefore be had at lower prices. This would contribute to offset the tendency toward higher average gas prices as a function of a greater proportion of Norwegian gas in the supply mix. This would be politically highly advantageous. However, such an outcome might require Norway to modify its gas pricing policy. Perhaps Norway would have to develop new institutional mechanisms, more suited for introducing a greater degree of economic, financial and industrial cooperation with the Western European gas buyers. In the absence of such changes the Soviet Union is likely to take the lion's share of the West European gas market. It will leave Norway as a fairly marginal supplier, at least in the short run. It is then only in the longer run, around the turn of the century, that Norway could become more significant by filling the place of the Dutch Groningen field as that finally phases out.

Notes

1. *Development of the Oil and Gas Resources of the United Kingdom, 1982*, London, 1982, Her Majesty's Stationary Office, pp. 4–5.
2. St.meld.nr. 40 (1982–83) *Om perspektivene bed petroleumsvirksomheten de naermeste årene*, Oslo 1982, Olje-og energi-departementet, p. 25.
3. Edward F. Denison, 'Economic Growth' in Richard E. Caves, ed., *Britain's Economic Prospects*, London, George Allen & Unwin, 1968, pp. 231–78.
4. Introduction to Caves, *Britain's Economic Prospects*, p. 5.
5. Anwar Jabarti, 'The Oil Crisis: A Producer's Dilemma' in *U.S. and World Energy Resources: Prospects and Priorities*, eds. Ragaei El Mallakh and Carl McGuire, Boulder, Col., ICEED, 1977, pp. 130–1.

6. Petroleumsvirksomhetens framtid, NOU 1983.
7. *World Energy Outlook*, Paris, OECD, 1982, p. 445.
8. 'Energie: désastre par abondance' in *Le Nouvel Observateur*, 22 July 1983, p. 30–1.
9. Burckhard Bergmann, *The European Market for Natural Gas*, paper presented at the European Conference in Oslo, Norway, June 1983, p. 15.
10. *Petroleum Intelligence Weekly*, 1 August 1982, p. 9.

Chapter 6

The International Energy Agency: problems and prospects

David J. Blair

The substantial reduction of oil imports by Western industrialized countries over the past few years has produced a feeling of relief in many corners. Some have even speculated that energy may cease to be a problem in the future. This sense of euphoria is nothing new. It was experienced in the period of 1975–8. At the time, the real price of oil was falling and supply was more than adequate to meet demand. Even in the early 1970s, there was similar optimism although oil production in the United States had begun to decline. These periods were abruptly ended by the well-known oil crises of 1973–4 and 1979 respectively. In both cases complacency had prevented oil-importing countries from making adequate preparations which could have saved them from most of the painful economic consequences of the Arab oil embargo and the Iranian revolution.

Collective action as preventive measure

While some of the fall in world oil consumption can be attributed to structural adjustment, in the form of conservation and switching to other energy sources, it appears that most of this decline is accounted for by the effects of the economic recession.[1] Hence, it can be expected that when economic growth picks up, oil imports will increase once more, shrinking the surplus that has existed in the oil market since 1980. As the Western economies emerge from the cycle of low economic growth, they will again find themselves in a position where they are vulnerable to even relatively minor disruptions in oil supply. If these countries are to avoid a recurrence of the dislocations of past oil shocks, they would be well advised to take advantage of the current respite by preparing themselves to deal effectively with future supply disruptions.[2]

One strategy for reducing oil import vulnerability emerged from the original oil crisis of 1973–4. It was that of collective action among the major oil-importing countries. The main vehicle for this approach was the emergency oil-sharing system of the International Energy Agency (IEA). Through this system shortages of oil would be spread among all member countries, rather than leaving one or two oil-importing countries to suffer the full effect of supply shortfalls on their own. Although the IEA has experienced a number of problems in its relatively short existence, its emergency arrangements remain potentially the most effective way for Western oil importers to reduce their vulnerability to supply interruptions. Before examining the problems and prospects of the organization, two of the main principles of the IEA's emergency system should be discussed. This seems particularly appropriate since their validity has not appeared entirely self-evident to a growing number of policy makers in recent years.

The first principle is the need for government intervention in oil markets in times of supply disruption. Government intervention in the economy has not been a terribly popular notion in a number of Western countries. However, there are a number of reasons why such involvement is a legitimate and necessary course of action. It can be argued that a free market has not really existed in petroleum trade at least since the Second World War, because prices as well as production, import and export levels have been controlled to one degree or another either by governments or oil companies. Today, the supply of oil in world trade is tightly controlled by government policy in oil-exporting countries. It would seem entirely reasonable for governments of oil-importing countries to exercise some degree of control over demand, at least in periods when supply is disrupted.

Oil and national security

The need for such government intervention is more apparent if it is recognized that oil import vulnerability is a national security threat. In order to appreciate fully the nature of the oil import vulnerability threat, it is vital to understand the multidimensional character of security.[3] That the security of nations involves more than just military threats was not lost on policy makers following the Second World War. The economic, political and social roots of armed conflict were recognized after the economic stagnation, financial chaos, protectionism, social strains and political extremism of the 1920s and 1930s. The non-military dimensions of security were uppermost in the minds of the founders of the United Nations. They insisted that economic and social issues be included in the organization's activities. The United States and its allies stressed economic development as a major security policy following the Second World War. American foreign economic policy during the post-war period—from the Marshall Plan to support for the establishment of the Common Market—was aimed at ensuring the economic prosperity and socio-political stability that were considered vital to the survival of liberal–democratic societies.[4]

The link between oil imports and the non-military dimensions of security is an important one because at first stance, oil supply disruptions do not seem to have much direct impact on military establishments. The defence sector is normally insulated from such disruptions by military stockpiles and by the priority it receives in domestic emergency allocations of oil supplies. Thus, oil import vulnerability is first and foremost a threat to economic security, and indirectly to socio-political security. Because oil plays such a vital role in Western economies, and because there are no readily available alternatives to imported oil, these countries are highly vulnerable to disruptions in the supply of petroleum. Supply disruptions can result in insufficient quantities of oil for domestic requirements and/or in sudden and significant increases in the price of the commodity. Either of these situations can have serious consequences for the economic welfare of Western industrialized countries.

The quadrupling of the price of oil which was induced by the Arab embargo of 1973–4 accelerated inflation rates, contributed to the recessionary trend, and posed balance of payments problems for Western countries. The economic recovery that was underway by 1978 was abruptly halted by a further doubling of the official price of oil resulting from the interruption of Iranian oil exports. Inflation rates were again out of control, recession set in, unemployment levels were the highest since the Great Depression and world financial markets were in danger of falling into disarray. Mounting economic problems such as these also increased the temptation of governments to adopt unilateral and protectionist policies, threatening the liberal international economic order, through which the economic prosperity of Western countries since the Second World War had been made possible.[5] Even those industrialized countries which were not net importers of oil could not escape the economic dislocations fostered by the oil supply disruptions, so long as they remained integrated in the world economy.

Although oil import vulnerability is primarily a threat to economic security, it can also be a direct or indirect threat to other dimensions of security. Oil supply disruptions may touch off intense competition for scarce supplies of petroleum among military allies thus straining alliance solidarity. The cohesion of alliances may also be weakened as members adopt such policies as import restrictions and export subsidies, aimed at reducing the economic damage induced by oil supply disruptions.[6] Beyond the problems that are created for military alliances, oil import vulnerability may adversely affect military security in individual states when economic recession reduces the revenue base of governments and forces them to consider cuts in defence spending.

The economic consequences of oil supply disruptions can also have a serious impact on the socio-political security of a state. Lester Brown points out that

> Inflationary stresses can quickly aggravate social divisions, turning political cracks into fissures. Perhaps the ultimate threat of uncontrolled inflation is that it eventually undermines public confidence in governments and institutions and can thus pave the way for violent shifts to the radical right or left. . . . it is becoming clear that expanding unemployment constitutes of the world's gravest social ills. As unemployment levels climb, the distribution of income within a society invariably worsens and further aggravates social inequities and political stresses.[7]

The ability of such economic ills to intensify social tensions and foster political extremism was clearly demonstrated in the interwar years. Elements of this pattern have also appeared over the past decade in many Western countries.

In view of the serious security implications of oil import vulnerability, the need for government action during supply disruptions seems clear.[8] Allowing market forces to prevail may be desirable when oil supply is adequate to meet demand, but the social costs of permitting uncontrolled price bidding when supplies are temporarily disrupted are simply too high. Government intervention in the economy is considered legitimate whenever particular commercial interests do not coincide with the general public interest, in this case, the interest of society's security. The oil market situation during the Iranian revolution is a clear example of such a clash of interest. Oil companies had reduced their oil inventories to dangerously low levels during the second half of 1978. Since there did not seem to be much likelihood of an increase in oil prices at that time, it was not profitable for the companies to hold on to extra oil stocks. If governments had intervened to pressure companies to maintain a higher level of inventories, or if governments had created sizeable strategic petroleum reserves on their own, much of the economic damage which followed the 1979 supply disruptions may have been avoided.

When Iranian exports were curtailed, oil companies scrambled for any available supplies on world oil markets. In many cases they built up inventories far above expected needs. They did so partly out of fear that supplies would not be available in future but also partly out of a desire to increase profits by stocking up in anticipation of further price increases. This, of course, was to a large extent a self-fulfilling prophecy, since the effect of stocking up in a tight supply situation was to drive prices upward. Again, if governments had acted at this time to pressure companies to refrain from building up their stocks, oil prices could have been prevented from rising as much as they did. The 1973–4 disruption resulted in a 7 per cent shortfall in world oil supplies. The disruption of 1979 was more limited in size. It reached only 5 per cent, and yet it had almost as severe an impact on the world economy as the previous crisis. The next oil supply shortfall may be much greater than the past two, but even if it is smaller its economic impact could be far more damaging. With the ability of the international monetary and trade systems and of domestic political systems to absorb a third major oil shock seriously in doubt, government action to prevent such a situation should be of the highest priority.

The second principle on which the emergency arrangements of the IEA are based is the need for international cooperation among oil-importing countries. This principle stems from a recognition of the condition of interdependence which exists among these countries. When it comes to international oil trade, 'interdependence' is more than simply the rhetoric of politicians—it is a hard, cold fact. No oil-consuming country, large or small, can escape the impact of a sudden, major increase in the price of oil. No oil-importing country can rightly assume that its actions on the world oil market will not have any significant effect on that market. If a supply disruption occurs when oil markets are tight it may not take a very large purchase on international spot markets to touch off a bidding war.

Unless coordinated actions are taken to control the demand side of the oil market during supply disruptions, countries risk taking unilateral measures which may harm other oil-importing countries, and ultimately themselves. The past experience of oil importers with such policies as pursuing special bilateral supply deals with oil exporters has shown the relative ineffectiveness of independent action in reducing oil import vulnerability.[9] Carrying out some of the measures that have been advocated by the IEA during supply disruptions without international coordination may also be ineffectual. For example, if a particular government persuades oil companies to draw down oil inventories, or begin to sell off its own strategic stockpile, its actions will be offset if companies in other countries are permitted to build up their inventories.

Policies aimed at limiting or ending dependence on oil imports are like throwing out the baby with the bathwater. Dependence on oil imports is not necessarily a negative thing, as long as a state can effectively reduce its vulnerability to external manipulations of the supply and price of imported oil. With effective policies in place for managing supply disruptions, the cost of import dependence shifts from the consequences of periodic supply disruptions (economic recession, inflation, domestic unrest) to the much lower cost of managing disruptions (through such measures as temporary demand restraint and the building of oil stockpiles). In short, if oil-importing countries are willing to prepare themselves to act decisively and collectively in future supply disruptions, they can continue to enjoy the benefits of imported oil, which is still a relatively cheap energy source, at an acceptable cost. While in theory this solution to the problem of oil import vulnerability is quite logical, in practice it has not been an easy one to implement. The following discussion will look at the original goals of the IEA, its past performances and some of the problems involved in organizing a collective response to oil import vulnerability.

Aims and performance of the IEA

The effort to organize Western oil-importing countries following the Arab oil embargo was an American initiative intended to serve a number of American foreign policy interests. The American government wanted to seize upon the issue of the energy crisis in order to reassert American leadership in the West and to buttress Western alliance solidarity. There was also a recognition of the political damage that the rapidly rising oil prices could wreak on Western economies and on the international liberal economic order. A major goal, then, was to end the competition among major oil-importing countries for supplies through bilateral deals with OPEC countries which was driving up the world price of oil. The United States also wanted to head off a possible shift in the foreign policies of Western countries toward the Arab side of the Middle East dispute, since this would complicate American mediation efforts in the region. Finally, it was hoped that a common consumer stance would lessen the West's vulnerability to pressure from oil-exporting countries, and prevent any further embargoes.

Governments of other oil-importing countries were initially wary of American overtures for Western energy cooperation. They feared that any moves in this direction would antagonize oil exporters and jeopardize their already uncertain supply situation. However, by the time of the Washington Energy Conference in February 1974, much of the tension created by the oil crisis had eased. Western European governments began to realize that their independent actions had not improved their positions in the oil market and that their competition for supplies had only succeeded in pushing up the price of oil. Most of them recognized the need for coordinated action and the need to cooperate with the United States which, as the largest importer and home of five of the seven major oil companies, was crucial to the success of any concerted consumer action. The Europeans and Japanese were also lured into negotiations on consumer cooperation by American offers of cooperation in the areas of alternative energy sources and energy research and development.

After months of negotiations sixteen OECD countries signed an agreement on an International Energy Programme (IEP) on 19 November 1974. It established the International Energy Agency as an autonomous body within the framework of the OECD. In addition to emergency preparations areas of concern of the IEA include: an information system on the international oil market and consultation with oil companies; long-term cooperation in reducing excessive dependence on oil; cooperation with oil-producing and other consuming countries; and coordination of research and development policies.

The central feature of the IEA is its emergency oil-sharing system. It provides a collective response by member countries to short-term emergency situations in which members suffer a petroleum supply shortfall. An emergency management manual setting out detailed rules and procedures for the sharing system was approved by IEA members in mid-1976. The system is supposed to be triggered when one or more member countries suffer an oil supply shortfall of 7 per cent or more, and involves demand restraint measures, the controlled run down of emergency reserves, and the reallocation of supplies available to the group.[10]

The emergency oil-sharing system is intended to serve three main purposes. First, the more or less unspoken aim of the sharing scheme is to deter oil-exporting countries from imposing any further embargoes on Western nations. It is with this implicit goal in mind that some observers have referred to the IEA as an 'economic NATO'. By banding together, Western oil importers hoped to prevent attempts by oil exporters to coerce them into altering their foreign and domestic policies.[11] A second purpose of the IEA emergency system is to reduce the friction among members of such institutions as NATO and the European Community during oil supply crises. Third, in the event that the deterrent failed, or if supplies were disrupted by some cause other than an embargo, the emergency system is intended to prevent the kind of chaos that was encountered during the 1973–4 crisis. Member countries would be assured under the plan of receiving at least 90 per cent of their normal level of supply, with the remaining portion to be made up by temporarily reducing oil consumption. It is hoped that this certainty would discourage oil importers from scrambling for supplies through bilateral deals or on the oil spot market, thereby bidding up world oil prices. In cases where only one or a few members were singled out for an embargo, the sharing scheme was intended to prevent economic dislocations which could end up affecting the entire international economic system.

The record of the IEA in meeting the aims of its founders has been somewhat mixed. While in some ways the agency seems to have played a very useful role in reducing the oil import vulnerability of its members, it also appears to have failed to fulfil its purpose in certain situations, particularly during the 1979 supply disruption. One of the problems involved in examining deterrence is that one can only be certain when a particular factor has failed to deter. However, it is not always entirely clear whether that factor is responsible for the target's restraint. Consequently, it is difficult to know to what extent the existence of the IEA oil-sharing system has prevented the recurrence of a major oil embargo. Other factors have probably been more significant, but the emergency scheme is certainly an important consideration which oil-exporting countries have to take into account if they are tempted to use that economic weapon. Unless the objective of oil-exporting countries is to wreak havoc on the world economy, the imposition of any future oil embargoes is not likely to be profitable for them, so long as the IEA system functions as it is intended to. Admittedly, this proviso is a rather large one. It points to the fact that the external credibility of the emergency system, that is its deterrent effect, is intimately tied to its internal credibility, or the confidence of its members in the system and in each other. As will be discussed later, there is good reason to expect that this mutual confidence will exist at least in serious supply shortages.

It is also difficult to determine to what extent the IEA has been responsible for the cohesiveness of such Western organizations as the European Community and NATO. Certainly instability in the supply and price of imported oil and the associated economic dislocations are likely to strain the organizations, particularly when much of the problem stems from the actions of other members. This is the case, for instance, when members attempt unilaterally to protect themselves by bidding up oil prices during supply disruptions. Insofar as the IEA has been successful in preventing this kind of reaction and has reduced supply and price instability, it has contributed to greater harmony among Western economic and security institutions.[12]

The IEA performance

For most IEA members, the primary purpose of the organization is to provide greater stability to the supply and price of oil in order to remove threats to their economic prosperity. In this role the effectiveness of the IEA is a bit easier to gauge than in its deterrence role or its role as promoter of Western solidarity. There have been two main tests of the IEA's effectivenss during supply disruptions since the Agency was established. The first of these was the 1979 disruption resulting from the revolution in Iran. The IEA was considered a failure by many after a relatively small interruption in supplies resulted in an oil price increase of over 150 per cent. It would be mistaken, however, to hold the Agency responsible for not preventing this oil price shock. The emergency oil-sharing system was designed to operate only during shortages of 7 per cent or more. The shortage during the Iranian revolution, however, reached only roughly 5 per cent. It may have been an error to set the trigger level so high. But in 1974, it was felt that some figure should be set so that the system would be activated automatically, unless the overwhelming majority of members opposed it. In a sense it was a case of preparing for the previous crisis, since the 7 per cent figure was based on the shortfall experienced during the Arab oil embargo. But there was also a feeling at the time that shortages below 7 per cent could be managed sufficiently by the major oil companies without any government involvement.

Multinational companies controlled about 78 per cent of world oil trade in 1974. By 1979 the figure had dropped to about 40 per cent, as oil-producing countries assumed greater control over their exports. This reduced the ability of the companies to reallocate oil during supply disruptions, as they had been able to do in 1973–4 without great difficulty. The 1979 supply disruption demonstrated that the market was no longer flexible enough to allocate supplies even in fairly limited shortages without prices being driven up significantly. Unfortunately, by the time this lesson was learned the damage had already been done. Once again panic had ruled the responses of a number of major oil-importing countries to a situation of supply uncertainty, and companies were permitted and even urged to bid higher prices on the oil spot market. Rather than being a case of organizational failure on the part of the IEA, the 1979 price spiral resulted from a combination of confusion within governments as to what was transpiring, and a lack of political will to act quickly and decisively along with other IEA members to bring the situation under control.

While the IEA was ill-equipped to deal with the Iranian disruption, a number of measures were gradually developed within the Agency during and after the crisis for managing future limited supply disruptions. One measure was the restraining of oil demand during limited shortages through the setting of oil import ceilings and targets. This tactic was employed during the Iranian crisis. By the time country-specific import targets had been set for all members the disruption had, however, passed and there was not much need for targets. While the setting of import ceilings is one possible way of dealing with limited supply disruption, it is probably not effective in relieving pressure on markets in the short term. In the past these targets were set on an annual basis, so there was not much incentive to reduce demand immediately since targets could be met by reducing demand at a later date. Also, import ceilings tended to be set fairly high so they were easily met. This limited their usefulness in crisis situations.[13] Another measure introduced in 1979 was the establishment of a registry of world oil spot market transactions, intended to prevent oil prices from spiralling out of control by bringing greater 'transparency' to the market. The usefulness of this measure was also limited, because most of the data were at least one month old when reported, and because only a small percentage of transactions was actually recorded.[14]

A measure which was to have much greater significance than either import targets or the spot market registry was the commitment made by the energy minister of IEA countries in December 1979 to meet promptly in the event of any major change in the world oil supply situation to decide on corrective action.[15] This commitment demonstrated that IEA members had become aware of their vulnerability to even relatively small oil supply interruptions and that they regarded government intervention during such disruptions as a necessity. Although the overall oil supply shortage in 1979 had only reached 5 per cent, the IEA was criticized for not activating the emergency sharing system when Sweden suffered an individual supply shortfall of over 7 per cent. This raised the whole issue of how serious shortages in only one or a few member countries should be dealt with. Under the IEP agreement shortages of 7 per cent or more in any participating country are sufficient to trigger the formal oil-sharing system. During the 1979 crisis, most IEA members shared the view that the formal system was a rather cumbersome mechanism to use for temporary, localized shortages. The Swedish shortfall had resulted from a combination of temporary technical problems and price controls. After the IEA Secretariat consulted with the Swedish government and with various oil companies, it was established that the problems facing Sweden, including transportation difficulties stemming from bad weather and refinery shutdowns for maintenance, were clearing up on their own. A decision was made not to trigger the emergency system since Sweden's supply situation was steadily improving, but also because of concern that activating the system might intensify the general crisis atmosphere and lead to more vigorous competition on the spot market.[16]

After the events of 1979 it became clear that the emergency sharing system was really intended to deal with large, generalized supply shortages. Shortly after the outbreak of the Iran–Iraq war in late 1980 a more informal approach to isolated shortages of over 7 per cent was developed. Turkey had lost about 70 per cent of her oil imports after the pipeline from Iraq was damaged. The IEA Secretariat consulted a number of oil companies in an effort to arrange for supplies to be sent to relieve Turkey's shortage. However, companies were asking for cash payment for the oil and at the going spot prices. A supply arrangement was held up due to Turkey's unwillingness to pay the price that was being asked. Eventually assistance became unnecessary when the pipeline reopened in January 1981. If the supply problem had continued much longer, however, the Turkish government would probably have had little choice but to accept the oil made available through the Secretariat.[17]

This informal approach to isolated supply shortages was endorsed at the December 1980 IEA ministerial meeting. The Ministers agreed that henceforth whenever a member suffered a relatively severe loss of oil supplies, the Secretariat would consult with the countries concerned and with oil companies. Taking these consultations into account, the Secretariat would then propose possible measures that could be taken by countries and companies to correct the imbalances and identify possible sources from which supplies could be obtained.[18] By coordinating the delivery of supplies to the hardest hit countries in a limited overall shortage, it was hoped that the Secretariat would be able to prevent sudden pressure on the spot market that could touch off a round of panic buying.

The commitment made in December 1979 by IEA members to meet promptly and to decide on corrective measures in the event of major changes in the oil supply situation was honoured in late 1980. Less than a week after the Iran–Iraq war began, and as spot market prices began to climb, an emergency meeting of the IEA Governing Board was held. Government representatives agreed to 'urge and guide' oil companies to refrain from making any abnormal spot market purchases and to absorb shortfalls by drawing down stocks held above the emergency reserve level required by the IEP agreement of ninety days of net imports. When energy ministers met in December 1980, they reaffirmed the decision to monitor closely the oil market situation, to continue the drawing down of stocks, and to encourage companies 'at an appropriately high level' to refrain from making unusual spot market purchases.[19] While ministers also agreed to consider the possible use of import ceilings, there seemed to be little enthusiasm for this measure. After 1980 the idea of import targets was dropped.

Through the winter 1980–1 the oil spot market stayed relatively calm. Oil companies drew down stocks at a rate of 2.5 million barrels a day, instead of the normal rate of 900,000 barrels a day. The market situation was helped by stepped-up production in certain OPEC countries, by the gradual return of exports from Iran and Iraq, and by a 6 per cent decline in world oil demand because of slower economic growth and conservation.[20] By March 1981 world oil production had surpassed demand, creating a glut on the world petroleum market. While many oil industry officials insisted that their decision to refrain from making spot market purchases and to draw on their record high oil stocks was taken purely on business grounds, the IEA did appear to have played a crucial role in maintaining market stability during the Iran–Iraq war.

In the opinion of a number of government and Secretariat officials, the IEA pronouncements and government consultations with oil companies shortly after the outbreak of the war helped to set a psychological climate that encouraged restraint on the spot market.[21] The importance of psychological forces in the world oil market was demonstrated in 1979, when the atmosphere of uncertainty grew into panic which led to a scramble for supplies.

The IEA's experimentation with these various demand management measures since 1979 demonstrates the utility of the organization in enhancing the oil import security of its members. The development of new ideas and responses could prove to be the most effective way of dealing with future disruptions, which may take on totally unexpected forms. Some of the measures developed in the past, such as import targets and the spot market registry, have proven not to be very effective. But this is to be expected whenever experimentation takes place. Other measures, such as the close monitoring of supply as soon as major changes appear on the oil market, the informal allocation of supplies to countries suffering significant shortages, and government pressure on oil companies to draw down stocks and exercise restraint on oil spot markets, may again be useful in managing limited supply disruptions. Different responses will be required depending on the circumstances and on the particular countries involved. The IEA provides a framework in which much needed innovation can take place.

Technical obstacles to crisis cooperation

The current problems facing the IEA's oil crisis management mechanisms are of both a technical and a political nature. Among the technical problems is the extreme complexity of the emergency sharing system. The system now involves over fifty companies and twenty-one countries. The growing number of restrictions written into supply contracts with oil-exporting countries has further complicated the task of directing the right grade of crude or type of product to markets that are experiencing a shortage. This problem of complexity has been dealt with to some extent by conducting periodic test runs of the emergency system. These test runs help the participants to identify difficulties that could arise in an actual emergency allocation and to work out solutions to these difficulties. In the latest test run conducted in the spring of 1983, for example, it was found that the introduction of a computer system had greatly facilitated the reallocation process.[22]

Another potential technical obstacle to IEA emergency arrangements is the lack of a mechanism for setting prices for oil which is reallocated in an emergency. This will probably not impede the operation of the formal oil-sharing system, where countries will likely be more concerned about obtaining supplies than about price. But the pricing issue has already posed problems during attempted informal allocations, such as the one arranged for Turkey in late 1980. Government pressure on oil companies may be required if reallocations of oil are not to be held up by pricing disputes in isolated shortages and in generalized 'sub-crisis' supply disruptions of less than 7 per cent. One of the factors that has hindered attempts to develop a means of heading off such pricing disputes is American anti-trust legislation. This has prevented the inclusion of oil prices in the emergency system test runs and has made American oil companies reluctant to cooperate with the IEA during limited supply disruptions.

The emergency oil-sharing scheme may also be hampered by inadequate emergency oil reserves. Including industry working stocks as emergency reserves means that much of the oil may not be usable in a crisis.[23] While this is not a serious obstacle during periods of oil surplus, as the supply situation tightens countries may find themselves short on reserves if they are required to use them.

A further technical problem with the IEA oil-sharing system is that in periods of declining oil demand the trigger mechanism becomes obsolete. The system is supposed to be activated when there is a shortfall of 7 per cent of the previous year's supply level. Under this rule, IEA countries were in a triggering situation during the period 1980–3, even though supply was more than adequate to meet demand. Thus, activation of the emergency oil-sharing system is not always automatic.[24] The triggering mechanism was considered one of the virtues of the IEA since it would prevent one or two countries from blocking its activation, as had been the case with the OECD sharing arrangements during the 1973 oil crisis. Also, a more or less automatic activation of the system would provide a degree of certainty about when emergency oil stocks would begin to be drawn down. This can be vital to maintaining market stability during supply disruptions. The trigger problem may become less serious with the development of measures to cope with limited supply disruptions. In the event of a clearly substantial supply shortage the chances of a consensus emerging among member countries for activation will be quite high. It may be difficult, however, for members to agree on activation of the system in situations somewhere between a relatively limited shortage in which *ad hoc* measures are becoming inadequate and a very large supply shortfall.

The most recent potential technical problem facing the IEA's crisis management is the level of readiness of its most important member, the United States. Under the IEP agreement each member of the agency is required to have ready a programme of contingent demand restraint measures to enable it to reduce its rate of oil consumption. In September 1981 the Reagan Administration allowed its domestic allocation programme legislation, the Emergency Petroleum Allocation Act, to expire. The Administration argued that in the event of a crisis rising prices would be sufficient to allocate petroleum supplies and to reduce demand. However, not only would this approach to allocation cause the more disadvantaged sectors of the population to suffer the most, but it would also rely on rising prices which is precisely one of the things the IEA was intended to prevent. According to the American approach, reduced demand would come *after* prices had risen, rather than being temporarily imposed in order to prevent a price increase which would cause undesirable economic dislocations. The Reagan Administration has argued that it possesses sufficient legal authority to allocate oil supplies if this becomes necessary. It has, however, refused to commit itself to any detailed measures for such an allocation beforehand. Also, the Administration has extolled the virtues of building high levels of oil stocks, but again has been reluctant to discuss in advance when and at what rate these stocks would be drawn down.

Without adequate plans for allocating oil and for using oil stocks much confusion could be encountered during a supply disruption, it may touch off yet another round of panic buying and spiralling oil prices. As David Deese has pointed out, 'If emergency measures are not in place before such a crisis strikes, we lose precious time and risk taking counterproductive actions. Many measures are only effective if they are taken before or at the start of a crisis'.[25]

The United States has adopted the same 'wait and see' approach in discussions within the IEA over the establishment of a mechanism for dealing with supply shortages of under 7 per cent, sometimes called 'subtrigger crises'. After the outbreak of war between Iran and

Iraq the IEA Secretariat and a number of participating countries supported the setting up of a mechanism for injecting oil supplies from surge stocks into the market whenever a supply disruption occurred in order to prevent prices from rising. The United States on the other hand rejected any formalized system for dealing with sub-trigger crises. It preferred to commit itself only to meet promptly with the other IEA members if an interruption in supplies occurred and to work out possible measures at that time. Again, the problem with this approach is that by the time agreement is reached on an appropriate response, assuming that any agreement is reached, a good deal of damage may already have been incurred.

The various technical problems outlined above could hinder the operation of IEA mechanisms for dealing with supply disruptions. But they will probably not be sufficient to prevent collective action from being taken as long as participating countries agree that such action is necessary.[26] The greatest threat to the collective management of oil supply disruptions is the loss of political will on the part of IEA countries.

Political problems

There are a number of potential political problems which are common to most attempts to coordinate the actions of states. First, differences in the capabilities of states may pose a problem for collective action. John Ruggie has observed that as national capabilities increase and become sufficient to perform a given task, the propensity for international organization decreases.[27] Thus, if a state considers that because of its economic strength it can afford to purchase oil on the spot market no matter how high prices are pushed, then it will be tempted to respond to a disruption of supplies by acting independently rather than cooperating with other oil-importing countries. Similarly, a country with considerable military capabilities at its disposal may, during a supply disruption, prefer a unilateral military response over collective action. After the experience of the two oil crises of the 1970s, the limited effectiveness and indeed the counter-productivity of such independent approaches seem to have been sufficiently clear to prevent countries from being lured by their economic or military strength into forgoing collective action for independent action. Even France, one of the most vigorous proponents of independent action during these two supply crises, has come to support the strategy of collective action.[28]

A second political obstacle to international cooperation can result from differing perceptions of vulnerability by various states. But again, the 1973–4 and 1979 crises demonstrated that all Western nations are vulnerable to the effects of even relatively limited supply disruptions, whether they are almost totally dependent on imported oil like Japan or virtually self-sufficient in oil like Great Britain.

Third, different conceptions of the state's role in the national economy can hinder collective management of international economic problems. This lies at the root of the apparent lack of adequate emergency preparation on the part of the American government, which has committed itself to reducing its role in the economy. It also explains the failure of IEA members to agree on a mechanism for dealing with subtrigger crises. Certain governments had regarded such a mechanism as undue interference in the oil market.

Although divergent attitudes toward state intervention may hold up collective action in low level crises, the IEA sharing system is likely to come into effect in more substantial emergencies. However, a fourth political obstacle could precipitate a breakdown in consumer cooperation. If it is believed that a crisis has been created unnecessarily by the actions of a particular IEA member, other members may be unwilling to share their oil supplies with it. An example of this is the embargo imposed by Iraq on oil exports to Canada after the Canadian government expressed its intention to move its diplomatic mission in Israel from Tel Aviv to Jerusalem. A number of Canada's IEA partners felt that this action was inspired by a desire to increase domestic political support for the government. Had the embargo spread to other oil-exporting countries, some IEA countries might have hesitated to provide oil for reallocation to Canada.[29]

A fifth potential political obstacle to the collective management of oil supply disruptions may arise if IEA members feel that they cannot trust one another to honour their commitments. Robert Jervis has argued that individual action may be taken, even if collective action is considered preferable when a state fears that other group members are unreliable.[30] Indeed, one of the greatest concerns about the unwillingness of the Reagan Administration to prepare adequately for supply disruptions is that this will create uncertainty among other IEA members over America's commitment to the IEA plan. Faced with a situation in which it is required to give up oil supplies to other members but lacking a pre-arranged programme for allocating supplies, it is feared that the United States may balk at activating the oil-sharing system. This uncertainty about American intentions could well cause a breaking of the ranks by those IEA members who anticipate a collapse of consumer cooperation during a supply crisis.

Doubts about the reliability of group partners can, therefore, threaten the success of collective action. But as Jervis argues, the gains of cooperation can be increased if each group member comes to value positively the well-being of the others in the group. This precondition does seem to exist among OECD countries. They have often demonstrated a rather clear understanding that the economic well-being of all Western countries is in the interest of each, although this understanding tends to become clouded from time to time. In order to avoid the breakdown of mutual trust among IEA members, steps need to be taken by the governments of member countries to demonstrate their commitment to collective action. This is one of the major reasons why many consider a pre-arranged programme for dealing with subtrigger crises to be so important.

A final political problem facing IEA emergency arrangements is that of domestic political opposition. It is perhaps the most serious of all. Public pressure on governments to withdraw from collective arrangements may be especially acute if consumers in a particular country are required to cut back their energy use by 7 or even 10 per cent, even though the country is not directly affected by a supply interruption. In countries that are required to redirect some oil imports or even domestic production to other IEA members, there could be strong public opposition to the idea of giving up oil supplies at the very time that a shortage is being experienced.[31] If a government is not in a strong political position at the time of a supply disruption it may be extremely difficult to resist domestic pressures to withdraw from the IEA oil-sharing plan. As Karl Kaiser has pointed out,

> it is difficult enough to get those elites which possess a certain expertise to understand the requirements of an interdependent system and to act accordingly, but it is even more difficult to extend that understanding to legislatures and to the public at large. How does one bring countries to understand that they are all in the same boat, and get governments or legislatures to act according to larger interests when they are elected on the basis of partial interests and are under pressure to solve their problems at the expense of other countries? The public will see primarily the immediate gains and not the losses to come.[32]

One of the most badly needed adjustments to the preparations of individual IEA countries is a programme of public information aimed at heading off domestic opposition to the sharing of oil. The public must now be made aware of the threat of oil supply disruptions to national security. It must understand that a coordinated response by Western oil-importing countries is the best way to protect the economic interests of the population. Such knowledge may alleviate domestic political pressure in times of crisis. Participation in the oil sharing system may also be more acceptable politically and cause less economic disruption if countries hold stocks in excess of the emergency reserve level required by the IEA. Instead of placing all of the burden of the 7 or 10 per cent reduction in oil demand on consumers, it would be preferable for governments to make up this demand restraint requirement by drawing down on excess or 'tertiary' stocks. It would gradually increase the rate of consumer conservation as these stocks are depleted.

Benefits and prescriptions

Despite the technical and political problems facing the IEA, there seems to be confidence within the Secretariat and among delegates to the IEA that the emergency sharing system can and will work in the event of another major oil supply disruption. This confidence may in itself be an important factor in making the system work. As long as member countries are aware of the benefits that can be derived from participating in IEA emergency arrangements it should be possible to assemble sufficient political will to overcome the potential obstacles.

What are the benefits that the IEA currently provides to its members? First, the deterrent effect of the emergency oil-sharing scheme may prevent the targeting of a member country for a selective embargo by oil-exporting countries. This provides greater freedom for oil-importing nations to formulate policies without fear of possible punishment from their oil suppliers.

Second, membership in the IEA provides countries with the assurance that oil consumers will face no more than a 10 per cent reduction in supplies, even if all of the country's imports have been cut off. Indeed, if governments opt for increasing their emergency reserves instead of imposing demand restraint measures, consumers may face no supply cutback at all. With their share of world oil trade declining, it has become more difficult for multinational oil companies to reallocate supplies in a supply disruption. Because the IEA emergency oil-sharing system includes oil imported by independent and state-owned oil companies, it can reallocate a much greater volume of oil than can the multinationals acting alone. Thus, the IEA offers protection for the portion of a country's oil supply that is not imported by multinational oil companies. If a member country suffers a serious shortage in relative isolation from other members, then the IEA Secretariat may be able, through consultation with governments and companies, to arrange for supplies to be delivered to that member.

Third, the IEA emergency management arrangements can provide greater stability in the prices which its members must pay for imported oil. Although the price effect will also be enjoyed by nonmembers, countries can increase the likelihood that prices will be stabilized by cooperating with other oil-importers since the larger the proportion of demand that is kept under control the more effective demand management will be.

Fourth, participating in the IEA emergency schemes gives countries the opportunity to have some influence on the policies of other members. It also allows them to have some input into devising new measures for demand management during supply disruptions.

The rationale for participating in IEA emergency arrangements may seem clear for countries that are almost entirely dependent on imports for their oil requirements (at least from the standpoint of supply). For countries that are more self-sufficient in oil supply due to domestic production, other benefits would seem more important.

This is true for the United States, Great Britain and Canada for which alliance solidarity, the reputation as a cooperative trading partner rather than being a free rider, are important motivations for remaining in the IEA.

In view of the benefits that the IEA offers its members, participating countries have a strong interest in ensuring that collective management of oil import vulnerability is effective. They could take a number of measures to increase the IEA's effectiveness. Thus, the United States could be encouraged to relax its anti-trust restrictions on oil companies so that pricing problems can be worked out during test runs and so companies will not hesitate to cooperate with the IEA in subtrigger crises. All participating countries should be encouraged to exercise their influence on oil companies to make oil available to members that are suffering a serious supply shortfall at prices that will not impede the allocation. A wider understanding should be fostered generally of the need for government intervention in oil markets during even low-level supply shortages in order to reduce the threats to the security of Western societies. Consideration should also be given to excluding industry working stocks from emergency reserves to ensure that all of these reserves can be used when needed.

IEA countries should make an effort to consult with other member countries whenever they propose to take some action which runs the risk of provoking a cut-off by oil supplies from oil-exporting countries. Such consultation may help the country concerned to discover alternative courses of action. Or it could convince other IEA members of the necessity of common action so that they will not refuse to cooperate in the event of an embargo. Measures should be taken by member countries, particularly the United States, to demonstrate their commitment to the collective management of supply disruptions by being fully prepared to carry out any measures that may be required under the emergency oil-sharing system.

In general, IEA countries need both to be adequately prepared for the kinds of emergencies that are anticipated, and flexible enough to adapt to unexpected crisis situations. Even if one member country considers itself to be adequately prepared for a supply crisis it cannot afford to be indifferent to the preparations of other members or to the effectiveness of measures that have been worked out to manage oil import vulnerability collectively. In future oil-supply disruptions the security of each Western nation will very much depend on the response of the others.

Notes

1. As Joseph Nye has pointed out, the problem of energy security has not been solved by the current oil glut. See Joseph S. Nye, Jr., 'Energy and Security in the 1980s', *World Politics*, **35** (October 1982), p. 125.
2. The inevitability of such disruptions has been accepted even by the Reagan Administration. For example, in testimony given before Congress a senior Department of State official stated, 'I fully agree with the point that if you look at the odds, the odds are that we are going to be suffering from some type of disruption over the next decade'. United States, Congress, Senate, Committee on Governmental Affairs, Subcommittee on Energy, Nuclear Proliferation and Government Processes, 'International Energy Agency and Global Energy Security Matters', 14 July 1981, p. 22.
3. Nye has defined security as the absence of acute threats to the minimal acceptable levels of the basic values which a people consider essential to its survival. Beyond the obvious value of physical survival, he identifies three clusters of basic values: a minimum expected level of economic welfare, group autonomy, and political status. J. S. Nye, 'Collective Economic Security', *International Affairs*, **50** (Autumn 1974), p. 588. For other discussions of the various dimensions of security, see: Lester R. Brown, *Redefining National Security* (Washington: Worldwatch Institute, 1977); Richard N. Cooper, 'Natural Resources and National Security', *Adelphi Papers*, No. 115 (Spring 1975).
4. C. Fred Bergsten, *Managing International Economic Interdependence: Selected Papers of C. Fred Bergsten, 1975–1976* (Lexington, Mass.: Lexington Books, 1977), p. 4.
5. For discussions of how major oil price increases have harmed Western economies, see amongst others: *Global Insecurity: A Strategy for Energy and Economic Renewal*, ed. Daniel Yergin and Martin Hillenbrand (Boston: Houghton Mifflin, 1982); Mason Willrich, *Energy and World Politics* (New York: Free Press, 1975), pp. 130–6.
6. The alliance strains produced by the 1973–4 and 1979 oil supply disruptions is discussed in Karl Kaiser, 'The Energy Problem and Alliance Systems', *Adelphi Papers*, No. 115 (Spring 1975).
7. Brown, pp. 33 and 36.
8. For a good study of how government intervention can achieve an economic optimum and why private market outcomes will fail to

achieve such an optimum, see Douglas R. Bohi and W. David Montgomery, *Oil Prices, Energy Security, and Import Policy* (Washington: Resources for the Future, Inc., 1982).

9. France's special supply arrangements with Iraq did not help that country during the Iran–Iraq conflict, and Japan's special arrangements with Iran did not improve her supply situation in 1979. For examinations of the experiences of some oil-importing countries with independent approaches to oil import vulnerability, see David J. Blair and Paul A. Summerville, *Oil Import Security: The Case of Japan and Great Britain* (Geneva: Programme for Strategic and International Security Studies, 1983).
10. For a more detailed discussion of the emergency oil-sharing system see: J. C. Woodlife, 'A New Dimension to International Co-operation: The OECD's International Energy Agreement', *International and Comparative Law Quarterly*, **24** (July 1975), 525–41.
11. This is an example of Klaus Knorr's conception of 'passive power', or the ability to limit what other countries can do to a country. See Klaus Knorr, *The Power of Nations: The Political Economy of International Relations* (New York: Basic Books, 1975), pp. 79–80.
12. See Wolfgang Hager, 'Western Europe: The Politics of Muddling Through', in *Oil, the Arab–Israeli Dispute, and the Industrial World: Horizons of Crisis*, ed. J. C. Hurewitz (Boulder: Westview, 1976), p. 42.
13. Interview with IEA official.
14. Interview with IEA official.
15. *IEA Press Release* (79), 29, 10 December 1979.
16. Daniel Badger and Robert Belgrave, *Oil Supply and Price: What Went Right in 1980?* (London: Policy Studies Institute, 1982), pp. 112–13; interview with IEA official.
17. Badger and Belgrave, p. 121; interview with IEA official.
18. *IEA Press Release* (80) 20, 9 December 1980.
19. *The Times* (London), 17 November 1980; *Petroleum Intelligence Weekly*, 6 October 1980; *IEA Press Release* (80) 20, 9 December 1980.
20. *New York Times*, 15 November 1980.
21. Edward L. Morse, 'Energy Security and International Preparedness', *Current Policy*, No. 272, United States Department of State, 23 March 1981, p. 3; interviews with delegates and IEA Secretariat officials; Badger and Belgrave, p. 124.
22. Interview with IEA official.
23. David A. Deese and Linda B. Miller, 'Western Europe' in *Energy and Security*, p. 193.
24. Activation is automatic to the extent that the system comes into operation once the Secretariat has made a finding that a seven per cent

reduction in supply has occurred or can reasonably be expected to occur, and if the activation is not overridden by a 'special majority' veto. See the *Agreement on an International Energy Programme*, Chapter IV.

25. Testimony before the Senate Subcommittee on Energy, Nuclear Proliferation, and Government Processes, 14 July 1981, p. 50.
26. This assessment is shared by some rather critical observers of the emergency preparations of the IEA. See Mason Willrich and Melvin A. Conant, 'The International Energy: An Interpretation and Assessment', *American Journal of International Law*, **71** (1977), p. 221.
27. John Gerard Ruggie, 'Collective Goods and Future International Collaboration', *American Political Science Review*, **66** (September 1972), p. 881.
28. Robert J. Leiber, 'Cohesion and Disruption in the Western Alliance', in *Global Insecurity*, pp. 337–8.
29. Interview with Canadian government official.
30. Robert Jervis, 'Cooperation under the Security Dilemma', *World Politics*, **30** (January 1978), pp. 167–214.
31. This was precisely the reaction of many segments of the Canadian public when press reports appeared in 1973 claiming that oil supplies were being diverted from Canada as oil companies allocated supplies among their customers. Public concern was also voiced in 1979 when oil companies repeated this action. See Blair, pp. 132 and 142.
32. Kaiser, p. 19.

Chapter 7
Options for European energy security

A. P. Bell

The central questions raised in this chapter are to what extent Europe must rely on imports and the degree of vulnerability to import supply interruption of the various sectors of its economy. Europe is neither richly endowed with coal, gas and oil, nor self-sufficient in nuclear fuel. However, the exporters of these fuels are not self-sufficient in their own strategic needs. All countries and regions of the world rely on each other for essential goods and services. As long as international trade continues unimpeded, our dependence on energy imports in itself does not threaten national or regional security. The problems that can arise from dependence occur when political factors interfere with the flow of commerce. Perhaps the best security we can hope for is that the basic laws of supply and demand will work to the benefit of both importers and exporters.

In my view, there are three aspects to the energy security problem in Europe. First, there are no realistic options for eliminating European energy imports by 1990. Second, we would require prompt and decisive action now in order to reduce imports significantly during the 1990s. Third, if we pursue a policy of deliberately curtailing imports we must be thoroughly aware of the impact that such a self-imposed embargo could have on the European and international economy.

Before examining some of the broader policy implications, I would like to take an overview of European energy demand and supply through 1990 and thus establish a range of European import requirements. With these data in hand we can assess the scope for reducing the level of European imports by 1990 and Europe's ability to withstand a short-term major reduction in imported oil or gas supplies. I will conclude with some brief remarks on the potential for changing the level of import dependence during the 1990s. All of the data I will discuss are for Western Europe. This excludes Turkey, Yugoslavia and the Comecon countries.

European primary energy demand

According to Esso's base case—a forecast that has much in common with other forecasts—total primary energy demand could rise from 1,175 million tonnes of oil equivalent (mtoe) in 1982 to 1,344 mtoe in 1990. This corresponds to an increase of 1.7 per cent per annum (see Figure 7.1).

Recent experience, however, has taught us to be very cautious about forecasts. It is possible to envision a range of 500 mtoe between an extremely low and extremely high demand for energy in 1990. Such a wide range makes it clear that we would be ill-advised to base our vision of future energy needs and problems within a narrow range of supply and demand. On the low side, demand could be about 1,000 mtoe in 1990 if there is little or no economic growth

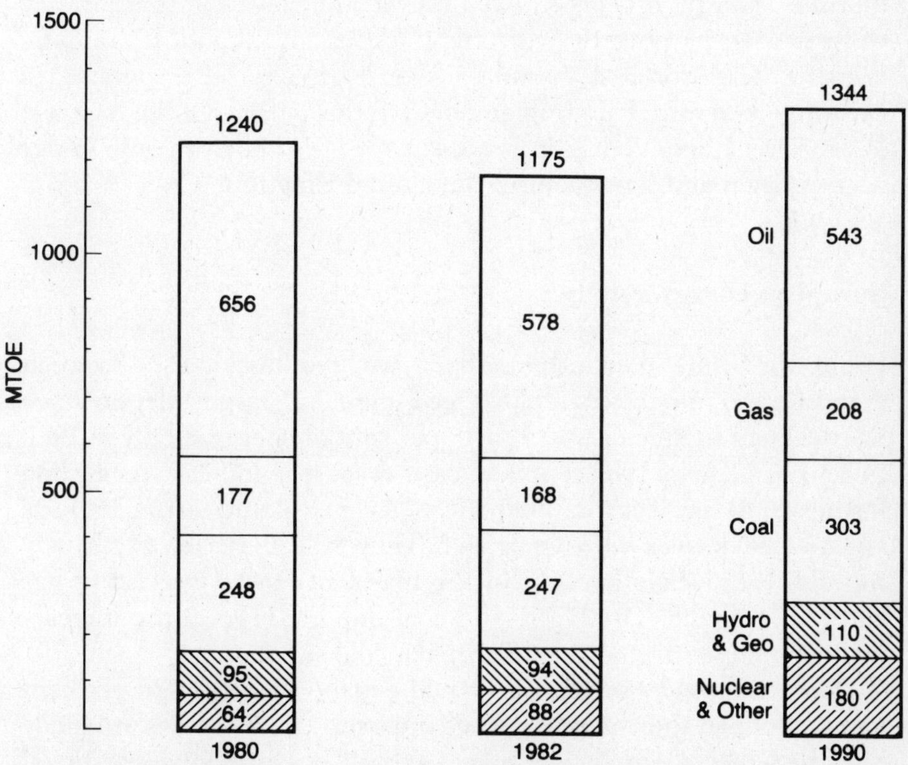

Figure 7.1 European primary energy demand.

and if crude oil prices are high. On the high side, demand could be about 1,500 mtoe in that year if we have economic growth above 3 per cent per annum and lower crude prices.

European oil demand hit its peak at 728 million tons (mt) in 1973 and then began to slide with a particularly strong decline after the 1979 oil crisis. Through the rest of the 1980s oil demand is likely to continue to fall because of the continuing move away from fuel oil. Natural gas demand reached a peak of 181 mtoe in 1979 and showed a slight drop thereafter. Unlike oil, however, growth is expected to resume in the early 1980s and continue through the 1990s because of greater use in the household and small consumer sector.

Coal demand is currently stagnant and likely to remain so for the next year or two. We do not expect real growth until the later 1980s and even then it will depend on a continuing price advantage over fuel oil and the installation of the necessary facilities for consumption. The small growth forecast in hydroelectric and geothermal energy reflects a lack of economic opportunities. By contrast, nuclear power is expected to experience the most dramatic growth. This forecast assumes that virtually all programmes currently committed are implemented in this period. Of the increase in capacity, 42 per cent is in France, 14 per cent in Germany, 16 per cent in Spain and 9 per cent in the United Kingdom.

European energy supply

Using the same demand numbers, we can distinguish between domestic and imported supplies (see Figure 7.2). Import dependence has declined significantly—from 49 per cent of energy supply in 1980 to 43 per cent in 1982. For our base case, it is forecast to decline further to 41 per cent in 1990. However, in volume terms, imports are now as low as we can expect. They will increase only slowly through the decade because of the major increases forecast in gas and coal demand. The big rise in coal imports reflects the increasingly uncompetitive costs of European coal supplies.

Domestic energy production could expand by nearly 20 per cent up to 1990 and thus help to reduce imports. The increases are likely to come largely from nuclear power and oil, but by 1990 domestic oil production is likely to fall from a peak of some 165 mt in the mid-

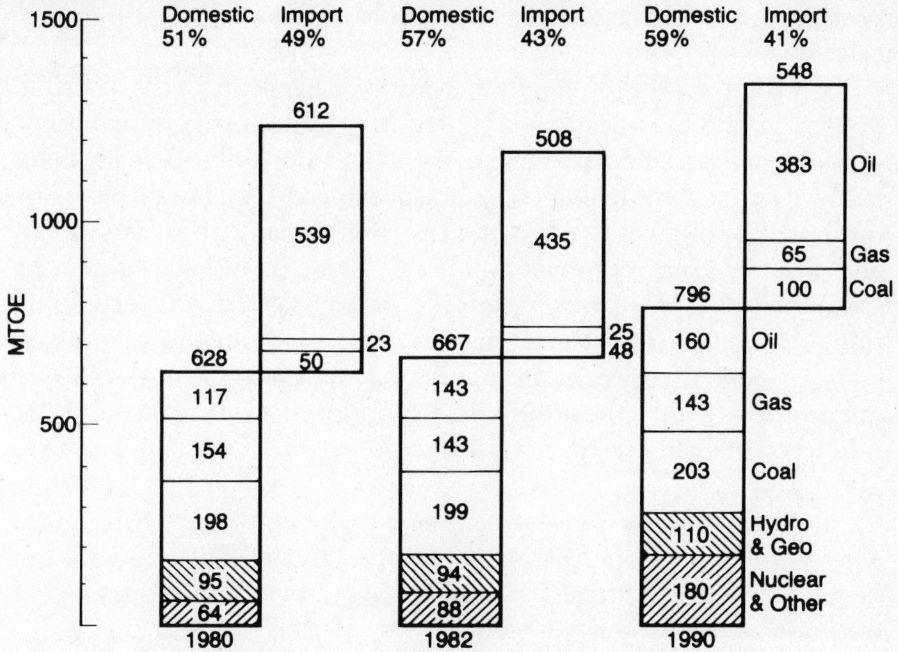

Figure 7.2 European energy supply.

1980s. Domestic production of gas has declined since 1980, partly because of the gradual depletion of fields such as the southern North Sea but mainly because domestic reserves have been shut out by imports. I will return to this subject later.

European energy imports: 1990 forecast

In 1990, Europe is forecast to import 383 mt of oil. This amounts to 71 per cent of total oil requirements and 28 per cent of total primary energy demand. In this time period, we are not too concerned about world supply capability. Europe's needs, in addition to comparable requirements for the rest of the world, suggest that OPEC countries will produce about 25 million barrels a day (mb/d). That is still below anticipated supply capacity but well above the 1984 level of 15 to 16 mb/d. Such an upswing in demand for OPEC oil is likely to stimulate some price increases. Furthermore, we face the political risks arising from the instability of some oil exporting nations, and possible international conflicts that could also disturb oil supply and prices.

An added source of concern in the years ahead is the possible impact on the European refining industry of importing much more product and intermediate feedstocks. This industry has already been badly damaged by falling demand, over capacity and surplus raw materials. By 1990, it is estimated that the capacity of North African and Middle East refineries will total 6 mb/d, twice as much as today's level. If this capacity is used in accordance with economic rules that differ from those characteristic of European refining, it could curtail investment in upgrading European refineries. The greatest risk is that these producing countries will look beyond the refining costs and use crude profits as an incentive to move products into Europe.

Natural gas imports in our base case make up 31 per cent of total supply and come from existing and proposed new contracts with the Soviet Union and Algeria. It is open to question whether this level of dependence is serious in view of the ideological and other differences between these countries and Western Europe. From a strictly commercial standpoint, the main impact of these Soviet and Algerian contracts in the 1980s will be to constrain domestic production in the Netherlands and the Federal Republic of Germany. Dutch and German supply contracts were struck in a period when natural gas was breaking into the European market. Hence they are considerably more flexible in their yearly and seasonal commitments than the recent Soviet and Algerian contracts.

Coal imports for 1990 in this case constitute 33 per cent of total coal requirements. Our information about world supplies indicates that they will continue to be abundant. New capacity will become available from several countries, including South Africa, Australia, the United States, Colombia and Canada. However, market penetration may be limited if European customers do not install the essential handling and burning facilities. Environmental concerns may also constrain coal's growth in Europe, and leave a higher demand for fuel oil.

In sum then, European imports of energy in any form are not threatened through 1990 by physical supply capability at the level required in our base case. If problems arise, they will be due to causes other than a lack of resources. We shall consider in the next section how Europe's energy needs would be supplied in two alternative cases: a 'high' case where demand in 1990 is 1,500 mtoe, and a 'low' case where 1990 demand is 1,250 mtoe.

European energy supply flexibility—1990: two scenarios

To determine the impact of a high demand case on individual fuels, we need to make certain assumptions about growth in the economy, and the pattern of consumption in each sector. A total demand of 1,500 mtoe in 1990 is consistent with a lower oil price and a GDP growth of over 3 per cent per annum from 1982. Of the total energy growth of 156 mtoe, oil demand could increase by about 57 mt. Much of that demand would be fuel oil for accelerated industrial activity. All of this increase would be supplied by imports. If similar economic growth occurs in the rest of the world, the pressure for higher oil prices would increase.

The increase in natural gas demand would amount to about 32 mtoe, and would provide an additional outlet for indigenous gas in Europe; Dutch and German production would be expected to increase to meet part of this new demand. Although stronger demand would stimulate new projects to develop European reserves, by 1990 they would not be on stream. So, imports would increase to about 85 mtoe or 35 per cent of total gas demand.

Coal could experience the biggest change with an estimated increase of 62 mtoe in demand. The rise could be even greater depending on the rate of installing coal-burning facilities. Almost all of the increase would come from imports. However, as mentioned earlier, there is ample world supply and keen price competition.

In total, an energy demand of 1500 mtoe would be expected to raise imports to 672 mtoe or 45 per cent of total demand. However, the decisive turnaround in economic growth from the recession of recent years would help to boost employment and improve the economic well-being of European citizens and governments. The higher demand case is good news even if it means an increase in energy imports.

Now let us consider how European energy requirements might be affected by a low demand case of 1,250 mtoe in 1990. This level of demand could be consistent with 1.5 per cent annual rate of increase in the European economy from 1983–90. Again, we have made some assumptions about the impact of such a low growth rate on each sector in order to assess the overall impact on each fuel. Under these assumptions, a fall in oil demand would diminish total demand by 43 mt. Fuel oil use bears the brunt because demand for steam raising and process heat declines in response to the low growth rate in GDP and industrial production. All of the reduction in oil demand would fall on imports.

Demand for natural gas in this low case would decrease by 18 mtoe. It would reduce not only production in Holland and the FRG but further discourage new developments in the North Sea. If one assumed that contracted gas imports were unaffected by such a drop in demand, additional indigenous supply would be shut out. Coal demand might also fall by about 23 mt and most of this decrease would fall on imports. But if even lower prices were available in world markets, there would be greater pressure to rationalize European production and reduce subsidies.

To sum up, in this low demand case the total share of energy imports would fall to 38 per cent of total demand and so help to reduce the impact that political events in other regions could have on the market. However, that slight increase in security would be the result of a very unhealthy economy with higher unemployment than already exists in today's very serious situation.

European energy-demand management

To this point I have discussed a considerable range of energy demand in 1990 and the effect of a high and low case on energy supplies, particularly on imports. I would now like to examine more closely how specific government policies could affect both demand and supply in Europe. First, let us consider the issue of direct government actions to manage energy demand.

From our observation and experience, the governments of Western Europe know more about the behaviour of energy markets than they did ten years ago. Most appreciate that market forces can support their energy policy goals of diversification and conservation and so they are less inclined to tinker with the situation than they were following the 1973 oil crisis. Several countries—notably Norway and the Netherlands–recently have joined the United Kingdom in abandoning oil-product price controls. Others, such as Italy and France, have relaxed their oil price regulations. We feel that the trend will continue to be in the direction of greater flexibility.

Nevertheless, some importing governments could be unwilling to rely primarily on market forces to manage demand as long as oil imports remain subject to politically inspired interruptions. Some may still feel that they should intervene to promote further reductions in oil demand through fuel substitution, excise taxes and the like, either at the national level or through the EEC. To assess the efficacy of that approach, however, let us review the impact on fuel consumption over the past ten years that government measures have had in comparison to higher energy prices. There is no doubt that higher energy prices have contributed to slow economic growth, particularly in energy intensive industries. In turn, recession and the painful restructuring of the European economy have depressed fuel demand. Since 1973 there has been a steady decline in the use of energy per unit of output in the economy and we would suggest that this trend will continue.

In contrast to the impact of higher prices and lower economic activity, direct government intiatives to reduce energy demand have had a much smaller and in our opinion not always productive effect. Exhortation through speeches and advertisements has had little influence on consumption. State subsidies for conversion from one fuel to another have proved to be a poor substitute for market forces because they have often encouraged uneconomic decisions. For example, incentives for industry to shift to coal have been based on the perception of significantly higher prices for oil or gas in the future. These price rises, however, have not materialized as yet. In a related area, some governments have provided a thinly veiled subsidy to local industry by offering artifically low electricity rates in certain areas. Such subsidies are more likely to interfere with, then to enhance, the economic use of all fuels.

We are all familiar with excise taxes on motor fuels. They have not risen in the last decade as fast as energy prices. For governments faced with significant financial problems they are bound to be an inviting opportunity if oil prices fall. Of all of the consuming sectors, however, the private transportation market appears to be the least elastic. The evidence indicates that demand is less responsive to higher prices in this sector and it is also of course very sensitive politically. Several European governments have increased excise taxes as oil prices have declined. To the extent that governments increase excise taxes to offset falling oil prices, they will be motivated primarily by the need for revenue. The tax rises will have little effect on demand.

In the near future, governments are likely to continue relying on market forces to manage demand for very pragmatic reasons. They have little scope in their national budgets to provide energy subsidies of any sort on a scale that would affect consumption. Above all, the primary goal of governments today is economic recovery. If governments replace market price reductions on industrial fuels with excise taxes designed to depress fuel use, such measures will raise fuel costs and so hamper Europe's competitiveness with the rest of the world.

European domestic energy supply potential until 1990

What can be done to increase the production of energy within Europe by 1990? In brief, not much. The key problem is lead time. Almost regardless of the need or the incentive, it takes about ten years from start to finish for a major energy project to come on stream, be it an offshore oil project or a nuclear power plant. There are a number of petroleum projects under development, in the North Sea, which will come on stream between now and 1990 (e.g. Hutton, Maureen, Alwyn, Clyde in the United Kingdom; Statfjord C, 34/10 and Vahall in Norway) but they will not prevent the downturn in production which we expect to begin in the late 1980s. The lead time required makes it very difficult to affect the 1990 production forecast significantly by policy changes or decisions between now and then.

The Netherlands have significant reserves of natural gas while Italy has small reserves which could contribute to the European energy supply in this decade. In both cases, government policy has been to keep these reserves for domestic use in later years. However, the Netherlands has indicated recently that it may allow additional exports in the 1990s. If the United Kingdom and Norway accelerated development offshore to the maximum (e.g. the British Southern

Basin and the Norwegian Ekofisk and 30/6 projects) they might produce as much as 8 mtoe additional gas supply by 1990. However, as long as gas supply capability exceeds anticipated demand, the business prospects are inhibiting.

Coal production in Europe might be increased by 5 mtoe but only at a very high cost. In our view, German lignite is probably the only sizeable supply source which can compete profitably against imports. As for nuclear power, regrettably, any variability in the supply of nuclear electricity by 1990 is downward. Like other forecasters, we have had to reduce our estimates of nuclear power every year because of political factors. For example, the Socialist Government in Spain has threatened to abort the completion of several plants. The continued opposition from environmentalists, uncertainty over electricity demand and the possibility of lower prices for fossil fuels, further limited the scope for increasing nuclear power supply in this decade.

Vulnerability to a six-month crisis

Having reviewed the various scenarios of supply, demand and import levels through 1990, it may be useful to examine how Europe would respond to a cut in its oil or gas imports emanating from an international crisis—a cut that would be large in size but temporary. How, for example, would Europe respond if it lost half of its oil and gas imports for six months? (See Table 7.1.)

Half of Europe's oil imports amounts to about 4.0 mb/d or about 35 per cent of the daily oil demand likely in the late 1980s. With this loss, the crisis supply programme of the International Energy Agency (IEA) could be triggered, and we would expect the following reactions (see Table 7.1). The first reaction is the implementation of the 10 per cent cut in demand specified in the IEA agreement for such a crisis, which amounts to about 1.0 mb/d. Most countries have standby allocation and rationing systems intended for just such an emergency but they will take several weeks to implement. The quickest way to achieve a major cut would be to replace fuel oil in power stations and major industrial plants wherever there is duel firing capability. However, this could turn into an unmanageable fuel oil glut and do little to alleviate a shortage of transport fuels. Any

Table 7.1 Vulnerability to six-month crisis

		mb/d
Oil	Assume 50% loss of imports	(4.0)
	Compensated by:	
	10% demand cut including substitution	1.0
	Drawdown of commercial inventories	0.5
	IEA supplies	0.0–2.5
		subtotal 1.5–4.0
	∴ drawdown of compulsory inventories	2.5–0.0
		bcf/d
Natural gas	Assume 50% loss of imports	(3.9)
	made up by:	
	10% demand cut including substitution	2.5
	Production increase	up to 1.8
	Drawdown of inventories	up to 2.2
Conclusion	Demand reductions have reduced bu not eliminated vulnerability to 'Europe only' crisis	

deliberate cut in demand would have to take into consideration the limited flexibility in refinery yields.

The second action listed is the drawdown of commercial inventories, that is any stock above the compulsory level. In recent months this has varied from five to fifteen days, and 0.5 mb/d reflects 10 days' inventory.

Supplies provided by other IEA countries could be as much as 2.5 mb/d if Europe alone lost half of its imports. On the other hand, if all the IEA countries lost half of their imports, Europe as a whole would not receive any extra supplies. There would of course be reallocation of available supplies within Europe. Both situations are extreme and arbitrary but they serve to illustrate the IEA system in operation.

These reactions either make up the full loss of 4.0 mb/d or only 1.5 mb/d, depending on the range of countries having their imports cut. Accordingly, we would have to draw on compulsory inventories by up to 2.5 mb/d. At the low level of inventories remaining after such a cut (about fifty days of supply), some parts of the system would probably have had stocks removed that are necessary to keep the system operating, and so more onerous consumer cutbacks would be needed. However, both personal welfare and the economy would be in reasonably good shape. The problem of course is that we would never know in advance how long such a crisis would last.

If Europe loses half of its gas imports for six months the response would be more straightforward and the situation less threatening. Half of those imports in the late 1980s is likely to be 3.9 billion cubic feet a day (bcf/d). To replace it, we have again assumed a 10 per cent reduction in total gas demand. This would provide 2.5 bcf/d. Again, the quickest way to do this would be to substitute fuel oil and coal in major dual-fired installations.

In such a crisis, increased production from Dutch and Italian reserves could provide more than the rest of the shortfall. However, there could be problems in distributing the gas or in the willingness of Dutch and Italian governments to bail out other European countries. With imports coming only from the Soviet Union and Algeria in this period, the loss would solely affect the countries that usually receive these imports.

Finally, to make up the loss, Europe could use emergency inventories established in France and Italy which are capable of providing 2.2 bcf/d for six months. Nevertheless, the French and Italian governments may not be entirely willing to come to the rescue of those countries that have not built up similar reserves. The many sources of supply for coal suggest that vulnerability to a cut in coal imports is insignificant.

In conclusion, it is apparent that lower energy demand would reduce Europe's vulnerability to supply crises. Still, a loss of either oil or gas supply in the proportions we have assumed would be a serious problem for Europe and a challenge to the relationships between 'have' and 'have not' countries.

Conclusions

There is clear evidence that the trend to lower energy consumption per unit of output is continuing. Lower prices will reduce the incentive for conservation and for investments to reduce energy use. However, we will continue to conserve through energy-saving practices and capital goods already as well as by the normal replacement of worn out plant and equipment.

We have also concluded that there is little scope for increasing the supply of energy in Europe by 1990. Increased oil production is not feasible because of the long lead time. The scope for gas increases is complicated by the current excess supply and new import contracts. Finally, any coal increase would be small and very costly.

Before concluding, however, that this situation is serious for Europe, it should be noted that the exporters of energy also have their problems. Nearly all the OPEC countries are in financial difficulties or need to dip into their financial reserves. The Soviet Union has a desperate need for hard currency. So all the exporting countries have a great need for our markets as sources of revenue.

Europe is expected to depend on imports for over 40 per cent of its energy supply in 1990. We probably cannot make do with much lower imports in this decade if the economy is healthy. Furthermore, we should not consider this level of imports to be a tremendous liability to our economic future or to our security as long as all nations can avoid a major political conflict that would interfere with energy and other trade. That, however, is a risk that countries have always lived with and no energy policy, however prudent, can entirely eliminate it.

Beyond 1990: opportunities and constraints

There are opportunities for significantly increased indigenous energy supplies in Europe in the 1990s if we take action soon enough. Looking first at oil, we are reasonably satisfied that the North Sea has a resource base sufficient to sustain the present level of production, say 3 mb/d through the end of the century. We are not optimistic about expanding reserves on the continent. But in the North Sea there are reserves, they can be developed at a cost within the present level of world crude prices and there is sufficient time to do the job. The constraints, however, are very real and serious. Fiscal regimes structured within the perspective of steady increases in real oil prices do not leave enough profit for the developer to proceed with many of these fields. The new tax regime for North Sea oil in the United Kingdom took significant steps in the direction of assisting development activity. But, if crude prices fall well below today's level and are perceived to stay low, many of these fields could still be too costly to develop.

The resource base for gas, particularly in Norway, presents even greater opportunities. However, offshore gas development requires costly pipeline systems and markets dedicated to these supplies by long-term contracts. The short-term gas supply/demand situation in Europe and the Soviet export potential cloud the prospects for these costly new North Sea projects. The high tax regime also inhibits potential development of these projects.

Europe does not have much opportunity to increase its domestic coal output economically. The current cost of subsidizing production from certain older mines tends to divert funds from the development of other mines that could compete with import supplies. But in the final analysis even if Europe does have to rely on some imports to supply the growing market for coal, this does not constitute a major risk.

Nuclear electricity remains an option for indigenous low-cost energy supply in the 1990s but on the whole not much progress has been made in this direction. The decline in the growth of demand for electricity has taken the spotlight off the need for nuclear power. Furthermore, even though its safety record is extremely impressive, sincere and active opposition from environmentalists cannot be wished away. In our view, its case needs updating and re-emphasizing.

My conclusion for the 1990s is that economic growth will require that energy demand continues to grow, though at a slower pace than before 1973. We will only be able to keep imports at the 1990 level if there are significant policy changes to encourage oil and gas production and major additional nuclear power programmes. Unless such changes occur very soon, the situation will deteriorate to such an extent that by 1993 it will be impossible to reduce imports before the end of the century.

Chapter 8

European energy security in focus

Ian Smart

Despite their joint popularity—or possibly because of it—neither 'energy' nor 'security' now conveys a sharply defined idea. When most people talk about energy, they mean fuel. Often, if pressed, they only mean oil. When most politicians or journalists speak, in an energy connection, of security, they mean self-sufficiency or independence, again with a heavy emphasis on oil. And they often turn out to be thinking only of dispensing with oil from OPEC or the Middle East. 'Energy security' has become a kind of code for all those fears and countervailing ambitions aroused by the experience of oil export restrictions and oil price increases in 1973–4 and 1979–80. And its revival during the more recent Atlantic argument over natural gas imports from the Soviet Union—themselves as an alternative, in part, to OPEC oil—has helped only to increase confusion.

Defining 'energy security'

Codes and private languages have their uses and charms, but they also have their dangers. Unless all use the same key, they endanger communication. More seriously, they too easily obstruct or distort policy. Undoubtedly, in present circumstances governments in Western Europe and many of their citizens have a strong, if vague, sense that their access to energy is at risk and an eagerness to reduce that risk. It is not clear that those who share that sense of alarm have a precise understanding of the problem they seek to address. Nor, therefore, can they easily judge whether the policies advocated to resolve the problem are well-founded or well-directed.

Ultimately, energy is matter and matter energy. Less rigorously, energy is a capacity to do work. Yet it has justly been said that no one wants energy. They only want what energy can provide: comfort, mobility and the myriad products and services in which energy is embodied. The point is not mere pedantry. Imagine, for example, a country that has one smelter large enough to meet all its domestic demand for aluminium by drawing on imported bauxite. The smelter, for reasons we need not explore, has hitherto been provided with energy from an amount of oil (or gas or coal) exactly equal to the country's total import of that fuel. In pursuit of energy security, the government closes the smelter and eliminates the country's dependence on imported oil. In consequence, the domestic market becomes completely dependent on aluminium imports from a single source abroad–possibly even the same source that formerly supplied the oil. Will the country or its economy, having sacrificed a share of added value, be more or less secure thereafter?

Nor is it only goods that simply embody energy which must be brought into the account. Those needed to extract or convert energy are even more obviously involved. Imagine now another country whose engineering industry is just able to manufacture conventional plant for generating electricity and all of whose electricity had hitherto been derived from imported coal or oil. In the interest of greater energy security, the government decides to substitute indigenous uranium, by turning to nuclear electricity. The corollary of reduced dependence on imports of raw fuel materials, however, is a new dependence on foreign sources for both the nuclear reactor equipment and the nuclear fuel fabrication then required. Will the outcome be greater or lesser energy security?

Setting aside for the moment the question of what makes anything more or less secure, those examples show that it is no more useful to consider energy security in isolation than it is to focus exclusively on fuel supply, as though energy or fuel were a final good. Energy is a means to some further end of economic or social value, as fuel is a means to the intermediate provision of energy. To act in ways intended to enhance fuel or energy security without any regard to other insecurities thus created is a prescription for disaster.

It should be no less evident that 'insecurity', as a code for some risk of supply being interrupted, represents only one of the factors to be weighed. Even within the artificially narrow context of energy, European fears are demonstrably related as much to the price of energy as to the physical availability of fuel. So they should be, since the crux of the matter is the total impact of meeting energy demand on the welfare of the economy or society concerned. However, one penalty of the fact that current apprehension has its roots in the experience with oil since 1973 has been a clouding of the true relationship between security of price (or cost) and security of physical supply. Reducing dependence on OPEC oil has seemed to evade that issue, by offering a defence against both embargoes and extortionate prices. Yet the distinction remains. A policy which substitutes expensive indigenous for cheaper imported fuel, or which relies on large investments to curtail the demand for imported fuel, may well increase the real price of energy to consumers. Conversely, a decision to continue importing relatively expensive fuel, while withholding a cheaper indigenous substitute as insurance against a future interruption of foreign supplies, may have the same result. Western Europe can offer examples of both. One benefit, in either case, may be partially to insulate the countries concerned against the immediate effect of a future price increase imposed by foreign suppliers or stimulated by some constriction of international supply. Meanwhile, however, an additional cost must be borne by those countries' energy consumers—directly, through higher prices, or indirectly, through subsidies from fiscal revenue. Some such premium may be amply justified. But how much more can be paid for, say, indigenous rather than imported coal in Britain and Germany, or for imported rather than indigenous gas or uranium in Britain or France, before the extra real cost begins to erode the security of the economy as a whole?

A parallel question arises with alternative energy imports. It may seem worth paying rather more for energy in the aggregate as a result of importing fuel from several foreign sources, rather than one, or of substituting imported gas or coal for imported oil. But how much more? And how is the calculation to be carried out if the question is one of substituting imported equipment or technology to extract or convert energy, at an ostensibly higher cost, for an alternative import of fuel itself? Finally, indeed, how are we to assess the costs and benefits, or even the real price effects, of importing finished products embodying a considerable amount of energy in their manufacture—say motor vehicles—instead of importing the fuel which we can convert into the energy needed to make the same products in our own countries? European governments have recently been much exercized by imports of motor vehicles from Japan. Yet each Japanese vehicle that displaces an indigenous alternative represents a reduction in Western Europe's dependence on imported energy resources. Are vehicle imports therefore to be viewed as a contribution to Western Europe's energy security? If so—and, by any narrow measure of energy security in terms of import dependence, it is so—then by what criterion, if any, is the cost reckoned to be excessive?

One implication of all this is that energy insecurity, together with any proposal for reducing it, must be examined in a wider context of economic and social utility. Another implication is that insecurity can only be assessed as a cost, the gross benefit of reducing which has to be set against the alternative costs of measures to that end. Some of those alternative costs can be expressed in familiar terms of money. Others, such as industrial employment, have a social as well as a monetary dimension. Yet others can only be measured in terms of security itself. It is significant, for instance, that imports of Soviet natural gas, originally seen as a more secure alternative to Middle Eastern and African oil, should since have been castigated as a new source of insecurity. That, too, is an alternative cost.

There would be less need to stress the wider context had European fears about energy security not come to be linked so tightly to dependence on fuel imports. In the first place, as should now be clear, fuel alone is a dangerously narrow focus. Exchanging a somewhat insecure supply of fuel for an even less secure or more expensive supply of energy-converting equipment or manufactures embodying energy may well be a bad bargain. In the second place, the implicit assumption that indigenous energy resources are perfectly secure is sadly ill-founded. The most obvious evidence to the contrary comes from the cases in which domestic supplies have, in fact, been interrupted. In 1973–4, for example, the cost to the British economy of a strike in the domestic coal-mining industry was considerably greater than the cost of the restrictions simultaneously imposed on oil exports by Arab members of OPEC. Similarly, the numerous costs of interruptions in electricity supply, as a result of industrial action, accident, plant failure or deficiencies of generating capacity, point to another limitation on the security value of indigenous energy systems in the European countries that have suffered them. And, in the third place, reliance on indigenous energy sources, as on imported fuels, always carries a cost. It may be negligible, or even negative. Yet there are assessments of the total social and economic effect of exploiting fuel resources in the Netherlands, Britain and Norway which argue otherwise. One may or may not accept those advertisements of economic distortion in Norway, 'deindustrialization' in Britain or the 'Dutch disease'. At the least, however, they underline the need to depict energy security as only one motif on a larger canvas. If a European country has fuel self-sufficiency within its grasp, but only at the cost of inflation, unemployment or greater dependence on foreign manufactures, will seizing the energy prize make it more or less secure?

The morals of the story begin to emerge. First, fuel security is not only about oil. Any such illusion should have been destroyed by the debate about Western Europe's natural gas imports, if by nothing else. Second, energy security is not only about fuel. Even in an artificially narrow context, the equipment and technology needed to exploit energy, together with the goods in which energy is embodied, must also be considered. Third, neither fuel nor energy security is only about imports. Pressing problems of security can and do arise in connection with the indigenous production, conversion or distribution of energy as well. Fourth, energy security is not only about physical supply. The prices to be paid and the real economic costs to be borne in meeting energy demand are as much a proper focus for concern about energy security as physical supply. Fifth, energy security, as commonly discussed, is a deplorably narrow concept, because energy itself is no more than a means to some further and larger end. Energy security is to be valued only to the extent that it conveys a net benefit in terms of economic and social welfare.

By following that circuitous route, we arrive at last at some idea of what energy security is, as opposed to what it is not. Security, in any connection, is about the relationship between values or assets, on the one hand, including the asset of freedom to shape future circumstances, and actual or potential threats, on the other hand. What we now see is that energy security, in Western Europe or elsewhere, is ultimately about paying a reasonable price for the freedom to pursue chosen economic and social objectives without being unreasonably constrained by threats to the physical supply or real price of energy. Patently, that begs as many questions as it answers. Some do not admit of a precise, still less a dispassionate, response. To select economic and social objectives, or to determine what prices or what constraints on the benefit of freedom are reasonable, is the very stuff of politics. And politics is both less and more than a precise, dispassionate process. Much of that political domain must here be considered out of bounds, not because it is political but because discussing it necessarily involves too much beyond the scope of energy policy. Two questions nevertheless remain. How does Western Europe view present threats to the physical supply and real price of its energy—and how should it view them? How can the constraints those threats represent be confined within limits, and at costs, which West Europeans might plausibly find reasonable? For an answer to the first question, at least, there is a good deal of evidence available.

Prescriptions for Europe's energy policy

Historical facts and figures about Western Europe's energy economies are well-described elsewhere in this volume. Future facts and figures are unknowable, but moderate projections over the short term between now and 1990 are likely to command fairly wide agreement. In 1983, 48 per cent of West European primary energy supply came from oil (71 per cent imported), 21 per cent from coal (18 per cent imported), 15 per cent from gas (21 per cent imported) and 16 per cent from hydraulic, geothermal and nuclear power. Imported oil therefore provided 34 per cent of all the region's primary energy, as against 4 per cent for coal imports and 3 per cent for gas imports.

The change from 1973 was already considerable. In that year, with Western European energy demand at roughly the same level as in 1983, oil imports made up 58 per cent of total supply, coal imports 3 per cent and gas imports less than 1 per cent. Albeit at a somewhat slower rate, future change is expected to follow the same trend. By 1900, Bell, in his contribution (see p. 131), suggests that Western Europe will still look to fuel imports for between 38 and 45 per cent of its primary energy, with a projection of about 41 per cent (which was exactly the level in 1983) as the 'base case'. Within that prospective total, however, the balance between different fuels will have continued to alter. Oil imports are expected to fall to only a little more than 28 per cent of primary energy—less than half the

1973 proportion. Gas imports will have grown dramatically, to reach 5 per cent of total energy supply. And coal imports, at over 7 per cent of total supply, will have more than doubled.

Without attaching ourselves to exact numbers, but recognizing that projections, like history, mirror policies as much as circumstances, we can apparently read between those lines the way Western Europeans have perceived the threats to their energy security, their options for responding to them, and the effective limits on some of those options. At first glance, for example, priority has clearly been, and still is, given to escaping from dependence on imported oil—by expanding indigenous production, reducing consumption and substituting other energy sources. Against a background of more or less constant energy demand, as between 1973 and 1983, that process has hitherto been accompanied by a comparable reduction in total dependence on imported fuels, only slightly mitigated by rises in coal and gas imports. During the remainder of the period to 1990, however, an expectation of slowly increasing energy demand entails an effective halt to the downward movement of total fuel imports, despite a continuing reduction in dependence on imported oil. In other words, coal and gas imports are expected to grow more rapidly, in order to balance not only the whole of the further reduction in oil imports but also a significant part of an increase in total demand. Even if demand were to remain almost static to 1990, however, limits on indigenous energy sources would be likely to mean larger gas and coal imports to off-set the persistent contraction of imported oil. With the future even more than the past in mind, the message is that dependence on imported oil, much more than on imported fuels in general, has been seen as a threat, to be reduced by demand restraint and an expansion of indigenous supply if possible, but by substituting other fuel imports if necessary.

Official projections to 1990 confirm, and even reinforce, that deduction. The International Energy Agency (IEA), for example, expects a somewhat larger total energy demand in Western Europe by 1990: 1,478–1,549 million tons of oil equivalent (mtoe), as compared with Bell's 'base case' of 1,344 mtoe.[1] Because its estimate of indigenous energy supply, although slightly more optimistic than Bell's, is broadly similar, one result is that oil imports are actually expected to be somewhat higher than in 1983, instead of falling by another 12 per cent. The other result, however, is that imports of coal and gas must rise much more rapidly, so that imported coal can meet over 9 per cent of total energy demand in 1990 and imported gas over 6 per cent. Thus, a major effect of more rapidly growing demand in Western Europe is a substantially larger absolute and relative dependence on importing fuels other than oil, presumably as a matter of policy.

The final confirmation of those deductions is to be found in explicit statements and decisions by West European (and other OECD) governments. The 'Principles for Energy Policy' adopted by the IEA Governing Board in October 1977 state that member countries have adopted as their overall objective 'reducing in absolute terms or limiting future oil imports through conservation of energy, expansion of indigenous energy sources and oil substitution'.[2] And IEA ministers, in May 1983, reiterated that 'dependence on imported oil ... remains the major risk to their [countries'] energy security'.[3]

So Western European leaders have formed a clear view of the one pre-eminent threat and of the means by which it is to be reduced. The official projections already quoted imply, in turn, some view of the probable limits on the efficacy of those means. 'Conservation of energy' and 'expansion of indigenous energy sources' having been pressed to their plausible maxima, there remains in 1990 a significant margin of demand to be met by 'oil substitution': that is, by definition, the larger import of other fuels. A simple security hierarchy is thus established. Conservation and indigenous supply are the most secure options. Non-oil imports are hardly less secure. Oil imports are indefinitely, if not absolutely, insecure. Or so it is made to seem. And the relegation of imported oil within that hierarchy is then underlined by adopting mandatory requirements, in the IEA and the European Community, for the stockpiling and emergency allocation of oil, but of no other fuel.

Before considering whether that basis for security policy stands up to scrutiny, it is worth looking a little more deeply into the analysis itself. Figures for energy supplies and imports, however accurate, can sometimes be misleading. IEA statistics indicate, for instance, that Western Europe's dependence on imported fuel is expected to fall from 61 per cent of primary energy in 1973 and 48 per cent in 1980 to only 43–45 per cent in 1990. Within those totals, oil imports ostensibly account for 58 per cent of all energy in 1973, 42 per cent in 1980 and 28–9 per cent in 1990, while imports of other fuels, on a similar basis, rise from about 3 per cent in 1973 to 6 per cent in 1980 and 15–16 per cent in 1990. What could be clearer? Given the evident conviction that energy security resides in dispensing with oil imports, what could more obviously justify a sense of modest achievement? Yet these and all the other unimpeachable figures for West European fuel imports quoted above need to be approached with caution, since they rest, in fact, on a special definition of 'import' and an idiosyncratic definition of 'fuel'.

In the first place, IEA figures treat any energy resource within Western Europe as indigenous to every country in the region. Thus, all the oil, gas and coal produced by Britain, Norway, the Netherlands and the Federal Republic of Germany is assumed to remain in Western Europe and to be immediately available to all Western European consumers, without entering into international trade (which is far from true). In the second place, IEA calculations regard all nuclear electricity as indigenous, without regard to the imports of fuel material on which about three-quarters of Western European nuclear power actually depends. In other words, the above estimates of import dependence are the artificial result of adding nuclear electricity to all other primary energy derived from resources found anywhere in Western Europe, and subtracting the answer from the region's total primary energy demand.

For our present purposes, the point is not so much the effect of those rather eccentric accounting practices on the advertised level of fuel import dependence in Western Europe. It may be interesting to note, in passing, that, when adjustments have been made for intra-regional trade and for all imports of nuclear fuel material, IEA projections actually indicate a need for Western European countries to import the same proportion of their fuels (59 per cent) in 1990 as in 1980, with the absolute quantity involved rising by some 150 mtoe. Here, however, what is more important is the additional light thrown on assumptions about the nature of Western Europe's energy security and the directions from which it is threatened.

We had previously discovered that indigenous energy resources, of any kind, are considered to be rather more secure than imports, and that all other imports are considered to be very much more secure than imports of oil. We now find, it seems, that fossil fuel imports from other Western European countries (or, possibly, from all other OECD countries) are reckoned to be more secure than comparable imports from elsewhere. Indeed, they are implicitly so secure that, together with imports of nuclear fuel materials, they are to be rated with indigenous production. It comes as no surprise thereafter to find also that, for security purposes, oil from non-OECD countries outside OPEC is preferred to OPEC oil, or that other OPEC oil is preferred to oil from the Gulf or the Arab world. That, too, reflects experience since 1973.

We have returned, of course, to where we began. To be more explicit, we have merely demonstrated the extent to which the current view of energy security in Western Europe has been shaped by experiences with oil in and since 1973. Only that can explain the pattern of assumptions underlying current policies and statistical projections. As a bonus, it can also help to explain some of the recent anguish in Europe and the United States over prospective Western European gas imports from the Soviet Union. Until the beginning of the 1980s, the reigning assumption was that any other means of meeting energy demand was preferable to importing OPEC oil, and that imports of coal or gas from anywhere, including any non-OECD source, were therefore to be welcomed without reserve. Governments were singularly ill-prepared, as a result, for the discovery that a non-oil import from a non-OPEC source might arouse fears about energy security, or that their respective views of the risks involved might then prove sharply divergent. Thus, it is only recently that the Atlantic argument over Soviet gas has resulted in the elements of a consensus on two new and important qualifications to the previous simple assumption: that some non-OECD gas imports are inherently less secure than others, and that security considerations argue for some quantitative, or at least proportional, limit on the amount of gas to be imported from a particular non-OECD supplier.

The second of those qualifications is especially significant, because it points to an element missing from the previous discussion of oil imports from OPEC. What worries people in Western Europe about the oil they import from OPEC members in general and the Middle East in particular is not only past experience of turbulence in its supply and price but also the simple fact that there is so much of it. In 1973, when 60 per cent of Western Europe's primary energy came from oil, imports from the Middle East alone provided 66 per cent of all the region's oil supply, to which African oil added another 22 per cent. In 1978, with oil still at 48 per cent of primary energy, the Middle East and Africa yielded some 75 per cent of the oil required. By 1983, the balance had shifted more considerably: only 48 per cent of Western European oil came from those sources (including 28 per cent from the Middle East), and oil as a whole met only 48 per cent of total energy demand. Between 1973 and 1983, therefore, imports of Middle Eastern oil into Western Europe fell from 40 per cent to 13 per cent of total energy, and African imports from 13 per cent to 10 per cent. Yet even those lower percentages continued to loom far above any other fuel import, and no less clearly above other major elements of primary energy supply. In 1983, for instance, imported Middle Eastern oil still outweighed the whole of Eastern Europe's gas production, while imported North African oil alone provided more energy than the region's nuclear power programmes. That quantitative preponderance is as much at the root of fears about the security of oil imports as any conviction that the physical availability or price of imported oil is at risk. Even Europeans who regard a major convulsion affecting Middle Eastern and North African oil as improbable are nervously aware of how costly it could prove were it nevertheless to occur.

A reaction to that nervous awareness is explicit in the Executive Director's preface to the IEA's 1982 *World Energy Outlook*:

> The energy challenge is clear: we must achieve a balanced distribution among energy resources. The alternative is to run the risk, through overdependence on one energy resource–oil–, that our economies will again be subject to sharp jolts as political events disrupt the international oil market.[4]

The emphasis there on the need for a 'balance' between energy sources and on the danger of 'overdependence' points to the last piece in the jigsaw of criteria by which Western European (and other OECD) governments have decided to assess energy security and to select means of defending or enhancing it. It is not only a matter of what *kinds* of fuel imports or what *sources* are insecure; it is also a question of *amounts*.

By deduction from this and other evidence, it does not, in fact, seem unfair to summarize the established Western European attitude to energy security in the form of a composite policy prescription:

1. Consumption of energy is to be reduced, through improved efficiency of conversion and use, to the lowest level compatible with not imposing a specific constraint on economic output.
2. Indigenous energy production, including all nuclear electricity, is to be preferred on security grounds to all other sources of supply.
3. Fossil fuel imports from Western European (and other OECD) countries are to be ranked, for security purposes, with indigenous production.
4. In meeting the remainder of demand—that is, in importing fossil fuels from non-European (or non-OECD) sources—the following principles are to be observed:
   (a) in the interest of reducing exposure to economic turbulence in the international market concerned, oil imports in general, and imports from OPEC countries in particular, are to be reduced;
   (b) in the interest of reducing exposure to political turbulence in the countries and regions concerned, oil imports from the Middle East and (North) Africa are to be minimized (and, if possible, eliminated);
   (c) to the extent necessary (and provided no other excessive exposure is thus created), oil imports from OPEC countries, and especially from the Middle East and (North) Africa, are to be replaced by alternative fossil fuel imports in the following order of preference:
      (i) coal or gas from any non-OECD source;
      (ii) oil from any non-OPEC source;
      (iii) oil from OPEC sources outside the Middle East and (North) Africa.
5. In case international oil supply should nevertheless be interrupted, emergency stocks are to be established and maintained at a specified level and, together with other available oil, are to be allocated on the basis of a predetermined formula if supply falls below a stated threshold.

The occasional bracketed words and phrases indicate areas of continuing uncertainty. In addition, there may be an element of persistent ambiguity—for example, as between the second and third rules. Overall, however, those rules have the superficial appearance of a precise directive. Indeed, it is a good deal more precise than anything Western European governments, jointly or separately, have placed on the public record. In the light of earlier discussion of energy security, it is nevertheless confusing, questionable and, in some respects, threadbare. It forces us back to first principles.

A threefold critique

It is easy enough to pick holes in those energy security rules. One hole has already been opened—and only partially closed again—by the dispute over the security of gas imports from the Soviet Union. That has yielded an acknowledgement that the process of substituting imported gas for OPEC oil must, after all, be limited by reference to the potential cost of 'overdependence' on a single source. Some will still want to ask, however, whether buying smaller quantities of gas from several OPEC sources is necessarily more secure than buying a similar total from a single Soviet source. Conversely, those who fear the spread of violent ideological conflict in and from Central America may now want to ask whether Latin American oil is clearly a secure alternative to oil from the Middle East. For that matter, is it clear that coal or uranium from southern Africa is as secure as the rules imply? Or that uranium from North America, which has been subject to political embargo more recently and for longer than Gulf oil, is nevertheless as secure as Western Europe's own oil, gas and coal? Is it even certain that domestic fuels in European countries can safely be omitted from a list of insecure resources?

Detailed challenges of that kind are far from trivial. They are less worrying, however, than the suspicion of more fundamental weaknesses. If the notional rules have been set out in a way that fairly summarizes current policy, then at least three basic criticisms might, in fact, be levelled at the attitude Western European states can be seen to adopt towards energy security: that it reflects too narrow an analysis of threats, that it concentrates on too narrow a range of options and, above all, that it displays an inadequate sense of how total costs and benefits need to be assessed when addressing the uncertain future.

Whenever we worry about security, we are concerned with two things: the probability that an undesired change will occur, and the cost that will be imposed on us if it does occur. As we have seen already in the case of Middle Eastern and North African oil, the level of our apprehension is the combined product of probability and potential cost. In the case of energy security, the undesired changes we contemplate are ostensibly of two kinds: substantial reductions in physical supply, and substantial alterations–normally, but not necessarily, upwards–in price. In the final analysis, price and supply, in relation to demand, are obviously linked. A reduction of supply, in isolation, imposes upwards pressure on prices. An increase of price, in isolation, tends to mobilize additional supplies (as well as to depress demand). Nevertheless, in practice and for policy purposes, some distinction between issues of supply and price is often useful, and sometimes unavoidable. A price increase may, for instance, have to be absorbed for some period, as an additional real cost, before supply can expand in response to it. Ultimately, however, neither analysis nor policy can safely ignore the link between the two. Nor is it generally prudent to ignore the links between the probability and cost of an undesired change.

When we consider responding to threats to energy security, it follows that we should be weighing the probability of the changes we fear, as well as their potential costs, and thinking of changes in price as well as supply. We should also be aware that probability, change and cost are all relative: in relation to some fixed baseline or to alternatives, but also in relation to time. As we know to our pain, a substantial increase in the price of oil may cause as much damage because of its abruptness as because of its size. Conversely, the probability of a particular kind of supply interruption may be higher three years hence than during the next six months.

The only way to extract some usable guide to action from that morass is to carry out a rigorous examination of all the costs and benefits involved, weighted to take account of different probabilities and time horizons. The penalty for anything less than rigour may include being caught unprepared by a change that could have been anticipated. But it may also include incurring an excessive cost for guarding against some contingency that is insufficiently likely or damaging to justify such expense. Worst of all, it may condemn us to discover too late that one risk has been provided against at the expense of creating or exacerbating another and more serious danger in the longer term.

None of that is novel or unfamiliar. Analysing the costs and benefits of security in any connection is notoriously difficult. We can measure the inputs, in terms of expenditure, but we cannot measure the outputs or benefits, because security is an imprecise concept and because there is no general consensus on its value–as anyone concerned with national defence budgets will ruefully testify. Almost by definition, moreover, we are forced to compare a certain, immediate cost with an uncertain and indefinitely postponed benefit, with all the attendant problems of deciding how small a probability and how remote a future time horizon demand our attention. Thus, in the present case, we are trying to deal much less with the current pressure of security factors on energy than with the apprehension of future changes in that pressure over an indefinite term. And it is not only the probability, nature and extent of possible change and the likelihood that, if it occurs, it will be reversible within a given time. In short, we are enmeshed in the problem of planning under multidimensional uncertainty. As a matter of common sense, however, we still have no choice but to apply the most rigorous rules of analysis we can. The temptation to be guided by ephemeral instinct or chance experience is, above all, to be resisted.

It is at that fundamental level that the most serious doubts may exist about the current perception of Western Europe's energy security problems and the appropriate responses to them. Too many of its features suggest that it may have been formed exclusively by an intuitive reaction to painful but limited experience, and that, as a result, both novel dangers and ulterior costs may have been overlooked. As we have seen, for example, the influence on analysis and policy of past difficulties with oil–the Arab oil embargo and the OPEC price rises of 1973–4, the Iranian revolution and the Iran–Iraq war–is all too evident. That has arguably meant an excessive preoccupation with oil supply, to the exclusion of security threats to other energy forms and sources. It has also meant an equal preoccupation with fuel, to the exclusion of other energy assets, and with imports, especially from areas historically subject to violent conflict, to the exclusion of security problems affecting actually or notionally indigenous supplies.

Striking evidence for one of those tendencies comes again from the case of Soviet gas. Had all eyes not been focused so tightly on OPEC oil, and had the consequential assumption that any other energy source was indefinitely more secure not been allowed to pass so long unchallenged, the issue of a prudent security limit on import dependence in that other context would plausibly have been addressed by and between the governments concerned before, rather than after, contracts were negotiated and investments committed. (As it was, by their own admission, IEA ministers first addressed the question of European gas imports in any detail at the May 1983 meeting of the Governing Board.) By the same token, systematic attention might then have been given, in good time, to the case for extending agreed rules on emergency reserves and allocation to cover gas, as well as oil. Whether the lesson has now been learned—for instance, to the point of ensuring that all the security implications of substituting other gas imports for additional Soviet supplies are considered before pursuing projects in Africa or the Gulf—remains to be seen. As in the past with oil, there is evidence of an instinctive reaction to an unanticipated difficulty over gas imports. But the instinct, as reflected in general exhortations, has still to be converted into a method for determining the relative security value of alternative options.

Even with oil imports, current security policies prompt some sceptical questions. One of their features, for instance, is that they are orientated primarily to a major interruption of supply. Looking to both the past and the future, however, it seems that the most serious costs may be those imposed by large, rapid changes in the price of oil, only loosely linked to the scale of reductions in supply. In 1973–4, in 1979–80 and again in 1981, conflict or political action in the Middle East had relatively small effects on the volumes of oil actually available to Western Europe. In none of those cases, in fact, is it clear that the supply reduction would have been large enough to trigger even the first stage of the emergency allocation arrangements now in place. On at least the first two occasions, however, a small supply change permitted or provoked a very large and abrupt price change—to the considerable detriment of most Western European economies. Yet agreed arrangements, or even basic rules, for dealing cooperatively with potentially disastrous price rises were notably lacking. Only in December 1981 did the IEA Governing

Board formulate the first elements of an agreed policy on 'sub-crisis' disruptions of the market.

The outstanding example of a cooperative security arrangement for dealing with large interruptions of oil supply is the IEA agreement on holding and using emergency stocks and allocating available supplies in such an event. In the light of experience, however, as well as with the future in mind, one might have thought it even more important to maintain stockpiles which could be drawn down to smooth the sort of drastic price changes we have actually faced, rather than to cope only with what seems to be a lower probability of massive supply interruptions. As a complement or an alternative, one might also have expected more serious consideration of how imaginative and flexible fiscal policy should contribute to the same end, by helping consumers to adjust to an abrupt change in world prices over a longer term. Such suggestions always attract charges of distorting economic reality. Although governments have, in fact, intervened in energy markets since time immemorial, all those objections to interfering with their play are familiar, and some are persuasive. Indeed, one of the strengths of the consensus already reached by IEA governments is its pervasive emphasis on the cost of opposing market forces. It is one thing to stand against the direction of change in a market, however. It is quite another to limit the unnecessary casualties inflicted by the sheer pace of change, before a new equilibrium can emerge. Even after a seasonal famine, food supply and demand are eventually re-balanced. But too many people sometimes starve in the interim.

Whether or not the size of past supply interruptions has been exaggerated in shaping European policies on energy security, the overall impression created by their heavy emphasis on oil imports is that of a preparation to re-fight the last war. It would be quite wrong to go to the other extreme, by dismissing the possibility of the last war recurring. The Middle East is indeed a turbulent region (although its turbulence has had less effect on oil supply than most people imagine), and any security policy must continue to have regard to that fact. It is worth bearing in mind, however, that those responsible for Europe's energy security failed to provide effectively in advance for the problems created by Middle Eastern turbulence in either 1973–4 or 1979–80, partly because neither episode conformed closely enough to previous experience. Again, it is also worth remembering that recrimination over importing Soviet gas into Western Europe was exacerbated by a failure sufficiently to foresee the fears it would arouse on security grounds. Preparing for a repetition of the last war must not divert attention from the probability that the next serious threat to Western European energy security will again be of an unfamiliar kind. Are we, for instance, clear what will be done in future if, for some reason, the flow to Europe of American coal or Algerian gas is suddenly interrupted—or a highly improbable but extremely serious accident forces the temporary closure of, say, 80 per cent of the region's nuclear generating capacity?

The new risks of minimizing old ones

Over and above the suspicion that energy security attitudes in Europe are still too narrowly linked to oil imports and to the past and, within that area, better adapted to disruptions of supply than of price, a more far-reaching criticism is that they have been overwhelmingly concerned to minimize the cost of future changes, with insufficient regard to the probability that damaging changes will occur. This is not to accuse those responsible of failing to assess, as best they can, the probability of a revolution in Saudi Arabia, a major accident on a North Sea production platform or a total strike in the West German coal industry (although some scepticism about the range of contingencies studied methodically might still be in order). It is rather to remark that only incidental attention seems to be paid to how energy security policies directed to minimizing the impact of one contingency may affect the possibility that other dangers will arise. Circumstances dictate policies. But policies also shape circumstances.

In some cases, steps to reduce the potential cost of a particular energy contingency may reduce its probability as well. If, for example, dependence on one supplier of fuel is brought below some threshold in the interest of limiting the potential cost of an interruption, a secondary effect may be to make it less attractive for the supplier to interrupt the flow, because his bargaining power has also been critically reduced. Indeed, just that argument has sometimes been used in support of measures to reduce dependence on OPEC or Arab oil. On occasion, it can also be used to justify cost-limiting action in one connection, on the grounds that it will make cost-imposing action less likely in another. Building an additional nuclear power reactor in Britain, for instance, may be justified not only by its effect on overall electricity costs but also on the grounds that it will limit the probability of substantial future increases in the price of British coal. Such fortunate ulterior effects are not, however, the product of some natural law. On other occasions, action to reduce the potential cost of disrupting supply or price may have just the reverse effect on probabilities. The energy supplier concerned may be provoked into exploiting his market power while he still has it: a phenomenon arguably to be observed in Western European coal industries, but conceivably relevant also to some oil or gas exporters. Or successfully limiting the cost of depending on one energy resource may generate a greater temptation to exploit market power on the part of other suppliers whose resources have been substituted or relatively revalued in the process: a consideration highly relevant, on the fact of it, to growing Western European dependence on imports of coal, or even uranium, as an alternative to oil. The moral is not that like causes produce like effects, but only that ulterior as well as proximate effects need to be weighed in the balance.

One peculiarly intriguing and important case in point is that of Western Europe's oil supplies from the Third World, with particular reference to the Gulf. The principal aims of energy security policies in the OECD and Europe has been to reduce dependence on that source. As a result of the success already achieved, two hitherto unfamiliar dangers begin, however, to emerge. One is a product of the fact that oil shipments, which mean energy to Europe, mean income and all that flows from it to the countries of the Gulf. In the absence of alternative markets, there is some level of oil income below which the probability of social and political conflict within Gulf states may rise quite rapidly. If the income loss appears to be unevenly or inequitably distributed between the different exporters—which is highly likely—the probability of international conflict between them may also rise. Given the large residual value of Gulf oil supplies to Europe, but given also the larger strategic costs of such conflicts, is it in the overall interest of Western European countries to reduce oil imports from the area indefinitely, without regard to the possible effect on those probabilities?

The other and related danger arising from efforts to eliminate dependence on Gulf oil should already be on Western Europe's agenda. In the relatively recent past, it was often said that the considerable dependence of the United States on the Gulf was to Western Europe's detriment: because American import demand helped to force up prices, or because American vulnerability to interference with Gulf oil supplies might inhibit the United States from responding decisively to Soviet action against Western Europe's larger security interests. Now, however, American dependence on Gulf oil has been reduced to a relatively trivial level, whereas Western Europe, as a whole, still relies on the Gulf for a substantial part of its supply. From that fact, there arises a fear that some Gulf oil exporters may feel less nervous about exploiting their remaining market power to political ends, because doing so will no longer seriously damage their relations with the United States (a significant reminder that commercial dependence can constrain sellers as well as buyers). And, as a corollary, there also arises an argument that Western European countries can no longer rely on the United States to defend their interests, as well as its own, in the Gulf, since those interests no longer coincide. In as far as either the fear or the argument reflects a naive assumption that American interests in the Gulf extend only to the American need for Gulf oil, it may not bear examination. Together, they remind us, however, that the effects of action to reduce the potential cost of an energy security contingency may not bear equally on the authors of the action, any more than on its targets.

That reflection also underlines one of the clearest and most durable lessons of experience since 1973: that disunity or selfishness among industrial countries can be as much a threat to their energy security as hostile action by fuel exporters. It is familiar ground, for example, that attempts by individual states to secure their oil supply by making exclusive government-to-government contracts with particular OPEC countries, except perhaps for marginal volumes, tend to reduce the security of other exporters by undermining the flexibility of the international market. In fact, they merely exacerbate the security problem already created by the loss of flexibility attendant on structural changes in the world oil market since the beginning of the 1970s: the loss of control over production by vertically integrated companies to host governments, the fragmentation of the oil trading system, and the rest. But lack of prudent coordination among Western European states can raise security issues in subtler ways than that.

To illustrate the point, an unfamiliar problem now arises as a result of efforts by some OPEC members—notably Kuwait, Venezuela and Saudi Arabia—to move 'downstream' in the oil market through the direct ownership of refining, distributing and marketing enterprises in industrial countries. On the one hand, that may be thought to reduce supply security, by tightening the dependence of certain European markets on specific sources of crude oil: in other words, by reducing flexibility. On the other hand, it may increase security by making it more likely that the OPEC exporters concerned will maintain supplies to those markets, even in time of trouble, in their own commercial interest. The more important point, however, may be that either effect will initially bear only on some European countries: on the Benelux countries, for example, in the case of Kuwaiti downstream movement, or on Italy in the case of Saudi expansion. As a result, if a serious problem of international oil supply were to arise in future, there could be friction between the principle of sharing the burden of stringency between Western European (and other OECD) states and the fact that those particular countries would somehow be in a special position—whether because they enjoyed a special advantage or suffered a special disadvantage. There seems little point in European governments trying to concert policy on downstream investment by oil exporters; national circumstances vary, and the potential costs and benefits are finely balanced. But there is certainly a point in discussing how such cases are to be regarded in the event of a future supply interruption, if only to reduce the risk of damaging recrimination at the time.

Whether the potential benefit of sustained supply outweighs the potential cost of diminished flexibility when non-OECD exporters own oil distribution facilities in Europe is only another version of the question about how the probability and cost of an energy security contingency affect each other. The same security trade-off between continuity and flexibility has to be assessed in considering the expansion of refining (and petrochemical production) within the territories of oil-exporting countries, notably in the Gulf. Will importing a higher proportion of refined products and a lower proportion of crude oil increase or reduce the security of West European oil supply? On the one hand, a major investment in export refineries should give Gulf exporters a larger economic interest in maintaining the international flow of their oil. On the other hand, the contraction of European refining capacity that seems likely to result, apart from actually reducing Western Europe's energy consumption (to recall an earlier example), might make it harder to switch to alternative oil if the flow from the Gulf were nevertheless interrupted. In any market, raising the cost to suppliers of withholding supply is calculated to lower the probability that supply will, in fact, be withheld. But there is always a countervailing price, and it would appear, in this case, to bear on the cost of consumers of that less probable, but still possible, event. How, then, is the balance to be assessed?

Although refinery capacity in Western Europe has already contracted considerably, the security problem to which that points may not yet be urgent. Looking to the future, however, European governments might be wise to consider now whether there is some lower limit on regional refining through-put, below which they should be ready to finance a security margin of reserve refinery capacity, as an emergency provision analogous to their oil stockpiles. As with emergency stocks, such a provision would visibly enhance the ability of Western Europe to withstand a future restriction of supply by particular exporters, and might thus reduce not only the cost but also the probability of such an interruption. Whatever the form of an energy emergency, we have learned that a readily available margin of flexibility is likely to be a particularly valuable tool for dealing with it.

Again, it would be wrong to consider the costs and benefits of flexibility only in regard to imported oil. Looking beyond the scope for reducing demand and exploiting genuinely indigenous resources, the most prevalent European reaction to fears about the security of oil imports has been to diversify along one or both of two axes: that is, to spread energy import demand across a larger range of fuels, and to spread oil import demand across a larger range of suppliers. Partly because an awareness of the distinction between those two kinds of diversification, with their different implications, was belated, there arose the danger of creating a new 'overdependence' on a few suppliers of alternative fuels, such as gas. Nevertheless, the instinct to seek flexibility through diversity was obviously reasonable, subject only to an assessment of its costs. The application should not be confined, however, to the structure of fuel import dependence. Taking a wider view of energy security, it applies with at least equal force to the whole field of energy conversion and distribution in Europe.

Towards a wider perspective of European energy policy

At the national level, sporadic but increasing attention has been given to the obvious security value of dual-firing for generating plant and industrial boilers. At the international level, IEA members have at last begun to look more carefully at the crucial importance of a more flexible, integrated regional pipeline system for distributing (or re-distributing) natural gas. But there is a long way yet to go. In the case of gas, for instance, the security value of a regional on-shore pipeline system will remain relatively limited while it continues to lack inter-connections with the major off-shore gas resources in the North Sea. No less strikingly, the security of Western Europe's electricity system will remain far lower than it might be unless much more is done to increase the flexibility of both national grids and their international integration. Because it reduces the aggregate requirement for costly spare generating capacity, more flexible electricity distribution is likely to be one of the most cost-effective long-term investments in energy security. That being so, it is remarkable that it seems to have received so little joint attention by Western European governments. Once more, the suspicion is not that policy-makers have identified the wrong objective. Security through structural flexibility, as pursued in diversifying oil imports, is clearly an admirable aim. The suspicion is rather that, by taking only a narrow view of energy security, they have pursued that objective in too narrow a context.

It will, of course, be objected that increasing the flexibility of energy conversion and distribution is financially expensive. So it is. Sometimes, as with the scope for saving investment in electricity generation, the economic rewards may also be very large. The fact remains that expenditure on greater flexibility still has to be justified by reference to all the costs and benefits, including the benefit of improved energy security. But it is as much a dangerous illusion to suppose that immediate financial costs automatically outweigh contingent security benefits as it is to suppose that greater energy security is worth any conceivable price. Calculating the present value of an investment in energy security is always bound to be a difficult and contentious activity, because we can offer only a probabilistic judgement on the value of the dividends and the rate at which they will accrue. But the immediate rules remain in force. With the benefit of security, as with any other benefit, nothing costs nothing—nor is anything worth an indefinite price.

One energy policy area in which those rules can too easily be forgotten is conservation. Just as current security attitudes and policies seem sometimes to suggest that OPEC oil imports represent a cost without benefits, so they sometimes imply that any reduction of energy demand, for a constant output, is a benefit without costs. Nothing could be further from the truth. There is nothing intrinsically good in avoiding the use of energy, any more than there is something instrinsically evil in using it. Setting aside purely spiritual and psychological rewards, energy conservation has to be justified in the same way as energy consumption, by reference to all its costs and benefits, whether measured in direct financial terms or in terms of security.

To illustrate that point, it is worth reflecting that reducing Western European demand for fuels is calculated eventually to reduce their prices, given a constant supply capacity. That conveys an immediate economic benefit to consumers. However, it also increases the economic pressure on higher-cost sources of current supply and on investment in higher-cost options for the future. If Western European countries were confident that the supply costs of their current and future indigenous energy resources are lower than those of any imported fuel, and that their security value is at least as high, that should not deter them. As it is, there is an obvious risk that, in the longer term, much lower levels of demand will actually erode the economic viability of indigenous energy in Europe, and perpetuate dependence on lower-cost imports. Saudi Arabia can produce oil at less than $5 a barrel. Australia can produce steam coal at less than $20 a ton. Western Europe can do neither—any more than it can match the production costs of Soviet gas or South African uranium. A careful balance is needed, therefore, between restraining demand and sustaining indigenous supply, and the proper balance can be struck only by assessing the costs and benefits of both courses in a context wide enough to embrace longer-term security considerations.

That may be appropriate as a final exhortation. Western Europe's energy will never be secure, because insecurity is in the nature of energy and the world. Even complete self-sufficiency cannot guarantee security to any of the states in the region. Apart from the security aspect of economic cost, risks of accident, industrial conflict and environmental prohibition always hang over indigenous supply. Nor is the argument that indigenous resources are at least secure *in extremis* a well-founded one; during two World Wars, European countries discovered that their need for fuel imports actually increased, because their domestic energy industries were so quickly disrupted. Yet imports are even more obviously insecure, being subject not only to accidental interruption but also to deliberate interference by foreign governments. Whether we think of Arab oil, Soviet gas, African coal or North American uranium, those threats are always with us. What changes with time is only the balance between the risks and costs of alternative ways to meet or regulate our energy requirements. The first task of energy policy, therefore, is to dissect that changing pattern, and to foresee its most probable future form. With the result as a basis for action, European governments have some chance of identifying the best and cheapest means to enhance, at any time, the total energy security of their countries.

Energy security is not a free good. Nor is its price as simple as we might wish. Its costs, as well as its benefits, need to be examined in a much wider perspective than has been common in Western Europe since 1973. The emotional shock, as well as the economic damage, inflicted by the experiences of 1973–4 and 1979–80 tempted Western European nations to abstract oil import security from energy security as a whole, while also divorcing energy security from the larger issues of economic and social welfare. At the same time, of course, it forced European governments to address a least a narrow segment of the problem more intensely and in closer concert than had ever been the case before—from which has flowed much that is good and useful. The hard truth, however, is that we are still only at the very beginning of addressing European energy security systematically. Carrying that necessary process further must now involve us in recapturing the wider vision of which we were deprived by a particular, urgent and largely unanticipated crisis ten years ago.

Notes

1. International Energy Agency, *World Energy Outlook* (Paris, 1982), Tables 10.30, 10.34, pp. 465, 469.
2. International Energy Agency, *op. cit.*, p. 46.
3. International Energy Agency, *Communiqué: Meeting of Governing Board at Ministerial Level: 8th May 1983* (IEA/PRESS (83)6), p. 4.
4. International Energy Agency, *World Energy Outlook*, p. 11.

Index